Environment and Agriculture: Rethinking
Development Issues for the 21st Century

Robert D. Havener
President, Winrock International, 1985-1993

Environment and Agriculture
Rethinking Development Issues for the 21st Century

Proceedings of a symposium in honor of
Robert D. Havener held May 5 and 6, 1993
at Winrock International, Morrilton, Arkansas

Winrock International Institute for Agricultural Development
1994

Copyright (c) 1994 by Winrock International Institute for Agricultural Development
Route 3, Box 376
Morrilton AR 72110-9537

All rights reserved
Printed in the United States of America

Library of Congress Cataloging-in-Publication Data

Environment and agriculture : rethinking development issues in the 21st century : proceedings of a symposium in honor of Robert D. Havener held May 5 and 6, 1993, at Winrock International, Morrilton, Arkansas.
    p.   cm.
  Edited by Steven Breth
  Includes bibliographical references
  ISBN 0-933595-85-9
    1. Agricultural resources—Congresses. 2. Agriculture—Environmental aspects—Congresses. 3. Agricultural conservation—Congresses. 4. Sustainable agriculture—Congresses. 5. Rural development—Congresses. I. Havener, Robert D., 1930- . II. Breth, Steven A. III. Winrock International Institute for Agricultural Development.
S401.R45 1994   94-1828
333.76'15—dc20   CIP

Acknowledgments

This book is the proceedings of a symposium in honor of Robert D. Havener, first president of Winrock International, and held at Winrock headquarters, Morrilton, Arkansas, May 5 and 6, 1993. The symposium was funded in part by the Winthrop Rockefeller Foundation, Ford Foundation, and Rockefeller Brothers Fund. It was organized by a committee consisting of John DeBoer, chair, Carol Stoney, David Seckler, and N. S. Peabody III. This book was edited and produced by Steven Breth.

Contents

1. The Havener Legacy in International Agricultural Development
Introduction, *Marion Burton*, 1

Agricultural Development and the Environment: Yesterday, Today, and Tomorrow, *Robert Havener*, 4

Panel, *H. Pat Peterson, William Dietel, Earl Kellogg*, 8

Discussion, 14

2. Conserving and Enhancing Soil Resources for the Future
Introduction, *Thurman Grove*, 23

Conditions for Achieving a Sustainable Global Agricultural System *Pierre Crosson*, 24

Managing the Living Soil for Human Well Being, *Richard Harwood*, 48

Panel, *Pierre Antoine, Paul Brown, Charles Fultz*, 59

Discussion, 64

3. Water Resources, Agriculture, and the Environment
Introduction, *Sandra Batie*, 69

Designing Water Resource Strategies for the 21st Century, *David Seckler*, 70

Panel, *E. Water Coward, Jr., N. S. Peabody III, Catherine Jewsbury*, 101

Discussion, 107

4. Poor People, Resources, and the Environment
Introduction, *Lowell Hardin*, 117

Political Sources of Agricultural Resource Degradation in Poor Countries *Robert Paarlberg*, 119

Panel, *Sandra Miller, Kenzo Hemmi, Neva Goodwin*, 126

Discussion, 132

5. Forest Resources and Forestry Policies
Introduction, *Roberto Rapera*, 139

Impact of Forest Resources and Forest Policies on Agricultural Productivity and Environmental Quality, *William Bentley, John C. Gordon, Charles R. Hatch*, 140

Panel, *Byron T. Edwards, Carol Stoney, Jim Wimberly*, 157

Discussion, 163

6. Range and Wildlife Resources

Introduction, *Fee Busby*, 173

Ecosystem Dynamics and Economic Development of African Rangelands: Theory, Ideology, Events, and Policy, *Jim Ellis*, 174

Panel, *Jim Maner, Ned Raun, Fanny Nyaribo-Roberts*, 186

Discussion, 190

7. Genetic Conservation and Biodiversity

Introduction, *Robert Horsch*, 197

Genetic Conservation and Biodiversity: Key Issues for the Future *Clive James and Avtar Kaul*, 199

Panel, *Jonathan Taylor, W. Ronnie Coffman, Robert O. Blake*, 226

Discussion, 231

8. Round Table Discussion: Implications for Development Assistance

Introduction, *Robert Havener*, 237

Major Issues for Discussion, *Vernon W. Ruttan*, 238

Round Table Discussion, *Robert Berg, Robert O. Blake, E. Walter Coward, Jr., Dan Martin, Michael Northrup, Peter Riggs, H. Pat Peterson, Elise Fiber Smith, Frank Tugwell*, 244

Closing Address, *Earl Kellogg*, 259

Participants and Authors, 263

1
The Havener Legacy in International Agricultural Development

Introduction

MARION BURTON
Board of Directors, Winrock International

The focus of this symposium is rural agricultural resources in the 21st century, environmental quality, natural resources, and technologies. The objectives are to (1) provide an overview of the state of the natural resource base relative to food requirements for the future, (2) highlight major environmental concerns as they will influence future agricultural production, and (3) assess emerging technologies and their potential impact on the resource base and the environment.

On this occasion I am privileged to reflect on the distinguished career of our retiring president, Robert D. Havener, and to make a few comments about his professional and personal accomplishments.

He was born on July 24, 1930 in London, Ohio. He cut his teeth on a crop and livestock farm in Summerfield, Ohio, and, being an ever-loyal citizen of Ohio, obtained a degree in agriculture at Ohio State University in 1952. After a 2-year stint in the Army, in an Army classified documents unit in Germany, Bob obtained a master of science degree in agriculture at Ohio State. In 1972, he rounded out his formal education by receiving a master's degree in public administration at the Kennedy School of Government at Harvard University.

Bob prepared himself well for the leadership role he has played over the years through work experience and education. He was a county extension agent and state extension specialist in animal science in Ohio while working on his master's degree. In his early professional life, he worked in many capacities related to agriculture, such as merchandising manager for two large U.S. meat processing firms, manager with the Ohio Farm Bureau Cooperative, and, in 1964, he joined the Michigan State University staff as a cooperative management advisor in Bangladesh.

Bob worked with the Ford Foundation for 12 years on a variety of programs internationally, such as managing agricultural grants to Pakistan during the Green Revolution, director of the Arid Lands Agricultural Development program, which focused on rainfed farming systems in 22 countries of the Middle East, and as the foundation's senior agricultural program officer for Asia and Pacific. In 1978 he became the director general of CIMMYT (International Maize and Wheat Improvement Center) in Mexico and served in that capacity until 1985. CIMMYT is a world-renowned agricultural research organization with more than 100 Ph.D. international scientists and administrators and 750 local staff conducting biological and social science research on maize and wheat, two of the world's most important crops. In all, he spent 18 years living and working in developing countries in Asia, Latin America, and the Middle East.

In 1985 we had the good fortune to convince him to be the founding president of the new Winrock International. As you know, that year Winrock International emerged from the merger of three not-for-profit organizations.

Bob Havener has received many honors: the Ohio State University Animal Science Hall of Fame, the College of Agriculture Distinguished Alumni Award, the American Agricultural Editors Association's Distinguished Service Award, and listing in *Who's Who in America*. Recently, the University of Arkansas recognized his accomplishments by awarding him an honorary doctor of laws degree.

He has served on the boards of numerous international organizations such as the International Center for Agricultural Research in Dry Areas, in Syria; International Rice Research Institute, in the Philippines; CIMMYT; International Agricultural Development Service in Washington, D.C.; Honduran Agricultural Research Foundation; the International Development Conference in New York; the Advisory Board of the World Food Prize; and the Ohio State University Agricultural Services Advisory Board.

As president of Winrock International, Bob took on the difficult task of molding the personnel and programs of three independent, nonprofit, agricultural development institutions into one. Combining three accounting systems is a task that will bring giants to their knees, just for starters. Through exemplary leadership and skill, he more than met the challenge. I can say with confidence that today we all think as one institution, Winrock International, which is dedicated to eliminating hunger and poverty in the world.

Bob has the unique ability to meet administrative demands, which could be all-consuming, and still provide vision and inspiration for program focus. In a rapidly changing world, he reacted in a most effective way despite restricted resources. He was able to focus on those areas of the world where the institution could have a beneficial impact and on the programs that could be most appropriate. In all of these activities, he consistently emphasized sustainability and environmental impact both internationally and domestically.

One accomplishment stands above all others in terms of vision and creativity. That is the African Women's Leadership Project. It has received universal acclaim and support. When Bob realized his friends and Winrock supporters wanted to honor him on his retirement, he requested that any honor benefit an important Winrock program. In response, his supporters created the Robert D. and Elizabeth Havener Fellowship for African Women Leaders in Agriculture and the Environment. I am pleased to report that $610,000 has been raised to endow the program. That's a fairly large statement of confidence and support. This is a fitting tribute, which will ensure that what Bob Havener has set in motion will continue.

This brings me to a matter dear to my heart and dear to many of you. Elizabeth Havener is Winrock International. Little did any of us know that we were getting two for the price of one, so to speak. As a matter of fact, one might be tempted to say, with great justification, that we'll reluctantly let Bob retire, however, Liz is indispensable and must stay. In all seriousness, I want to express our gratitude to Liz for the enthusiastic way in which she has worked with all of us over the years, for her very significant contributions, and for being the premiere gracious hostess.

With the departure of Bob and Liz, the Havener era will close and Winrock International will press on. But we will recall with great pride what was accomplished and we will reflect upon the way it used to be with our two good friends. There will be lasting remembrances and touches that even newcomers will sense as Havener influences. These proceedings are a symbol of the gratitude for what Bob and Liz have accomplished for Winrock International.

Agricultural Development and the Environment: Yesterday, Today, and Tomorrow

ROBERT D. HAVENER
Winrock International

One day in Peshawar, Pakistan, David Bell and I were talking. David, who was a mentor to many of us who were associated with the Ford Foundation, was there on a visit, and I was trying to impress on him how important well-trained people were in the development process. Traditionally the Ford Foundation had been mainly a grant-making organization, while the Rockefeller Foundation had been mainly a supplier of human talent to the development process. But in Pakistan we were following, more nearly, the Rockefeller model. And I said, "David, you don't understand. The transfer of capital assets to Pakistan is not the crucial issue. They have money to do what they really want to do. What they desperately need are very good people, and what the foundation ought to be doing is supplying first-class people who can understand what needs to be done, who know how to get it done, and who will set about doing it." David replied, "Well, maybe, Bob. I understand that Pakistan has some capital assets, and I understand that good people are important. But, have you ever tried doing it without the Ford Foundation's money?"

What I have been able to accomplish in the world is modest, and it was nice of Marion to try to make it sound important, but it has all been accomplished by working through other people and largely by using other people's money. It is indeed important that we keep such things in perspective.

Today the fundamental issue is whether or not we can build a brighter future for coming generations. In these unsettled times, that is no easy task. There are important changes occurring that will profoundly and detrimentally affect the future unless we find ways to accommodate them, to direct them, to control them, and to use them for the benefit of humankind.

During this symposium, we will focus on the soil and water resource base—the fundamental resources that underlie our agricultural activities and many of the other things that we do. We will spend some time talking about the genetic resource base of beneficial plants and animals from which we gain sustenance and enjoyment. We are also going to talk about the concerns for maintaining and enhancing our resource base for sustainable crop production, pasture production, and livestock and forage production. Clearly, both economic and demographic pressures are threatening the natural resource base on which future generations must rely.

The conference organizers asked me to reflect on the past 40 years or so that I have been involved in agriculture and rural development both abroad and in the United States and to highlight trends that will have a major impact on the next decade in international and U.S. domestic development. It may be useful to remind ourselves that environmental degradation is not new and it is not necessarily the product of high-input, high-production agricultural technology systems. The U.S. Dust Bowl of the 1930s certainly was not caused directly by high-input agriculture systems. The farm on which I grew up in southeastern Ohio certainly was a low-input farm, but, because of the poor resource base that we had, it was not a sustainable enterprise—at least it was not sustainable if I had to farm it. Thus, we should not assume that low-input agriculture is necessarily consistent with sustainability or that high-input agriculture is necessarily consistent with nonsustainability. Through all ages, soil and environmental degradation and regeneration has gone on. We have to examine the contexts within which inputs are used, the character of the inputs themselves, and the purposes for which they are being used. Our purpose should be to manage the agricultural resource base in ways that are beneficial to us and that will sustain that base for future generations so that they may use them to meet their needs at reasonable cost.

But what is different now than it used to be? Foremost is the population explosion. When I was born in 1930, there were about 2.2 billion people on this earth. Today, there are 5.2 billion. In the year 2000, there will be 6.2 billion. And by 2025 it appears that there will almost certainly be 8.2 billion people. To feed mankind, we will have to produce as much food and fiber in the next 35 years as has been produced from all agricultural endeavors since the beginning of human existence.

Yet, there is evidence that the productivity of major crops has ceased to increase at anything like the rates we have enjoyed for the last 25 years. I was recently in Colombia looking at research data from CIAT. That data showed that in Colombia rice yields still have not exceeded those of IR8, a variety that was released in 1965. In other words, on the best lands, using the best varieties, and with all the agricultural research, knowledge, and skills that can be brought to bear, maximum attainable yields have not changed significantly since 1965. Yes, new varieties are moving on to lands that earlier were not well-suited to rice production. They are moving there with more insect and disease resistance. But, under perfect conditions, they are at best equalling IR8 yields. In the Philippines, the story seems to be same for the new IRRI rice varieties. Despite what one hears about the new wheat varieties, this pattern also seems to hold true for wheat in Mexico. Nearly 30 years ago in Mexico, Siete Cerros 66 established a new yield standard, and until fairly recently, it was the highest yield that wheat researchers were able to achieve under ideal growing conditions. The crossing of the winter-wheat germplasm pool with the spring-wheat germplasm pool for wheat has resulted in an increase of around 10 percent because of heterosis, but, I believe that is basically a one-time event. In general, then, there are some worrisome signs of trouble on the genetic front concerning staple cereal crops and the prospects for raising their production potential on the land currently under cultivation.

There are also disturbing trends on the irrigation front. The rate of the installation of new irrigation systems has slowed, and in some regions of the world, irrigated area is declining because current irrigation systems are deteriorating faster than new systems are being constructed.

In many parts of the world, cities and towns are spreading into good farmland at an alarming rate. The spread often seems imperceptible, but when continued over a decade or over a generation, millions of hectares disappear as land is captured for nonagricultural use. The present concerns over the environment will not go away. They will be increasingly important on the development agenda. And they should be. That is why Winrock, early on, began to incorporate environmental considerations more carefully into the activities that it undertook. But, those environmental concerns have caused a paradigm shift in the price of production resources. Think about the price of timber. The spotted owl has done more to increase the price of lumber in the United States in the last 18 months than one could have imagined. Lumber prices have risen by nearly 100 percent in a matter of 18 months. I think that this new level will not be permanent, but that shows how greatly one environmental issue, the spotted owl, or the reaction to it by parts of the timber industry, can affect national and, indeed, international prices—and the costs of building and furnishing new homes.

In California, preservation of salmon runs and of duck and geese breeding grounds in tidewater areas have suddenly exercised a take-out bid on the price of water. Price itself is no longer the issue, water is being allocated based on politics—on public demand regardless of price. It is within that context that we need to consider new policy issues that we have largely ignored in the past.

The Earth Summit in Rio de Janeiro brought us to a new era. Agenda 21 is on the table, and it clearly will affect what the United States and the European governments do as donors. In many ways, the Rio conference was a very substantial coming together of genuine concerns about the environment. But as one looks into who will do what, when, and, more important, who will pay for the needed changes, there is much yet to be settled. But the issues raised at Rio will now color the development dialogue in the World Bank, the Asian Development Bank, the Inter-American Development Bank, and government offices around the world as we talk about the cost, returns, and proper approaches to development.

I happen to believe that on balance technology is still our friend and that development of new technologies will continue to be vital. In the early 1960s, triage was a legitimate discussion in development circles. People actually sat around in rooms like this and talked about how many people ought to be allowed to starve in India or Pakistan then, so that more people would not starve in India and Pakistan now. And the fact that such discussions seem odd now that we have had several decades without major crop failures demonstrates the power of technology when appropriately applied to potentially productive agricultural resource bases.

I was startled at a recent conference in England to hear Bangladesh cited as a success story in development. When I go to Bangladesh, I have trouble seeing it as a success story, but maybe it is. When my family and I moved to East Pakistan, now Bangladesh, in 1964, it was a country of 53 million people and it was importing 3.5 million tons of grain. In 1993, with 115 million citizens, Bangladesh has been declared to be self-sufficient in foodgrains. It has had an agricultural productivity increase of about 6 percent a year for the last 30 years. There are few countries in the world that have achieved that kind of sustained agriculture growth. While we still see many poor and hungry people there, that achievement is testimony to the rewards of properly applied agricultural technology.

At Winrock, we have looked carefully at the development process over the last 7 or 8 years. We reexamined our mission statement and asked, in this changing world, what should Winrock be doing? We modified the words that made up our mission statement. We used to say that we were created to overcome the problems of hunger through agriculture development. As a result of that strategic planning exercise, we added "poverty," recognizing that employment and poverty are at least as important in many places of the world as are total food supplies themselves. We also added the concept of sustainable development, recognizing that only if we had concern for maintaining the environment would we address the important long-term development issues. We realized that, if we truly were going to affect rural residents, there were things that were as important as, or in some cases more important than, agriculture itself—the environment in which rural people live and the rural infrastructure that supports them.

We looked at the important issues in development and said, there are four pillars of development: human resource development, institutional development, policy development, and technology development. I believe those four pillars are essential to sustained economic development whether urban or rural. But the most important of them is human resource development.

I am convinced that human resource development is the single most important tool that we have to enhance general human welfare and that Governor Winthrop Rockefeller's concern for helping people to help themselves should be our guiding touchstone in most of the things that we do. Our aim should be to help individuals and societies evolve in ways that are compatible with their traditions, culture, and aspirations, but in ways that open and expand the opportunities available to them. It was in that vein that Elizabeth and I chose the fund to support the African women's program as the thing that we wanted to feature out of issues that have been of importance to us. We are extremely grateful to those who have contributed to this fellowship fund established in our names. To me, this program represents a new version of the Agriculture Development Council and honors the vision of John D. Rockefeller III, who created that very successful human resource development program. But it also recognizes the extremely difficult development problems that face us in Africa and the role women must rightly play in overcoming those problems.

Finally, during these closing years of the 20th century, I think we must be very concerned about events taking place in Central Asia. Unfortunately we may look back on the Cold War as being the good old days, when there was a predictability and stability in that potentially volatile region of the world. It seems to me that we cannot ignore developments there or neglect trying to influence them in ways that are socially compatible with the goals of their people and with those of the rest of the world as well.

Panel

H. PAT PETERSON
U.S. Agency for International Development

To me, agriculture does three simple things: it feeds people, provides income and employment, and uses natural resources. And how it does that is a key to solving the population and food issue, which is more than just a population-food equation because income, employment, and natural resources, feed into the equation. It is an inescapable conclusion that there are more people now than there were yesterday, and there will be more tomorrow than there are today, and they are going to have to be fed.

That leads us to what I'd like to call the civilization-type issues that we're struggling with. One is irrigation because without large-scale irrigation we will not be able to get there, and irrigation systems are deteriorating. We have to solve that problem. Second, I don't think we've put enough effort into dealing with rainfed agriculture. That's why Africa suffers again and again from droughts and famine. We haven't had technical breakthroughs in rainfed agriculture comparable to those we've achieved in irrigated agriculture. A third issue on the horizon, which isn't really a technical agriculture issue but directly relates to what we're dealing with, is the epidemic of AIDS. Over the next 20 or 30 years, it will affect how many people we have and how they relate to agriculture. AIDS takes out of the productive process those people who are economically viable. In Africa, we're already seeing a shift in the dependency ratio that's moving people from high-value agricultural production back into subsistence farming because of the labor constraint.

Finally, there's one more civilization-type issue, at least in the short term. We must do something about the restoration of depleted natural resources especially in the Newly Independent States of the former Soviet Union. We cannot afford to have the second and fourth largest nuclear powers in the world suffering from huge unemployment, triple-digit inflation, and food scarcity. Russia and Kazakhstan are both dependent upon agriculture as the basis for their productivity, which is threatened by the deteriorating natural resource base. We can't let that situation continue. At least it makes me uncomfortable thinking that Kazakhstan is out there with all its missiles pointing toward me.

There are some things I believe that we can do. There are technologies around that would help us examine some of these issues. There's a growing awareness that we need to manage natural resources and not just protect them.

I believe that there's a lot more we can do with computer technology through simulations to accelerate research activities. There's a lot we can do with computer technology to manage the vast amounts of data we need for working with large-scale irrigation systems. Breakthroughs in communications will help us with that. After all, 50 years ago, if you were at the headwaters of an irrigation system, you had to put somebody on horseback who would have had to ride all the way down to the end and come back before you knew what was happening. Now we can find out instantaneously.

I think that the real breakthrough in agricultural technology, the next Green Revolution, is going to come through biotechnology and through genetic engineering. We're now on the brink of making some major changes in our technology.

Finally, we need to change the way we do agricultural research. We need to move more away from monocropping systems and go to multicropping systems. And we will have to start taking a much stronger look at mining of the micronutrients in the soil. That seems to be one of the key reasons we're losing productivity with rice in South and Southeast Asia.

WILLIAM DIETEL
Board of Directors, Winrock International

Several points that Bob touched on deserve amplification. The evidence appears to be overwhelming that we have to develop some new styles of institutions to help carry out the changes that are coming because of population pressure and changes in technology. It's also clear that the communications revolution of recent decades has raised the level of expectations of the poor and of large parts of the world's population who heretofore have been denied access to participation. And they will demand participation in the process of change to a degree that none of us could have imagined 20 years ago.

It's also clear that some of the great institutions of the past, many of which are represented here, are going to be insufficient for the tasks that this changing world will demand. They're going to have to find new partners, and they're going to have to recognize that they can't carry out their mission unless they accept the facts of interdependence and reliance upon other institutions with which they've had little to do in the past.

The growth of the nonprofit sector globally is one of the biggest secrets of our time. None of us really know, today, how many of these institutions exist. Through some preliminary studies around the globe, we're beginning to get a better sense of their proliferation. But the magnitude of it, the rapidity with which they are coming into being, and the extent of their impact is still beyond our grasp.

This month in Barcelona, there will come into being a new organization, for the moment called Civicus. It is an effort to bind together, on a global basis, nonprofit organizations, donors, and recipients. And quite understandably that organization is initially going to touch down at the ground level. It will find, I suspect, its influence and its impact not among the big institutions but among the small ones—the grassroots community organizations that are proliferating at such a fast rate.

This will transform the institutional world in which we operate. It will make organizations like Winrock International, the Ford Foundation, all kinds of entities, both on the donor and donee side, rethink how we're going to work collectively to create some of the changes that must occur if we're not going to end up with a global disaster.

Some organizations, and I hope Winrock will be one of them, will continue to find the resources, both human and financial, to help in exploring new ways of creating institutions and of sustaining new institutions and new institutional styles that we don't know very well, but that are growing at a rapid rate. That will be part and parcel of that four-part mission that Bob Havener talked about for Winrock International.

Let me make a personal comment by way of conclusion. I couldn't help think, as I listened to Bob, that if John D. Rockefeller III, had been permitted to have the length of life to which he was certainly entitled, how pleased he would have been to be here today to hear about the African women's program. Before his death, this man, who was one of the earliest of our countrymen to see the population problem coming and who was the founder of the Agricultural Development Council, one of the great human resource development agencies, saw the importance of the role of women, and I think he would have seen the importance that they are playing in creating those institutions at the ground level that I just mentioned. And how pleased he would be that the Agricultural Development Council is now part of a larger organization and has been led for the past 8 years by a man with a wonderful wife, both of whom are committed to human resource development and particularly to the role of women. Surely that is our great ace card: the possible employment and involvement of women in solving the problems we have been talking about.

EARL KELLOGG
Winrock International

I have given some thought to demographic changes during the span of Bob Havener's career because these changes are critically important to agricultural and environmental development. When he began his professional career, there were 3 billion people in the world, and we had a population growth rate of about 2.3 percent, adding about 69 million people every year to that stock. World population is now 5.5 billion. Population growth has dropped considerably, to about 1.7 percent a year, but we are adding, with a larger stock, about 93 million people each year. Based on the medium variant for the United Nations projections, we'll have about 700 million more people by the year 2000.

Accompanying this significant population growth, urbanization will proceed rapidly. Nevertheless over 50 percent of Africa and Asia's population will continue to be rural even in 2020. And in developing countries as a whole, rural populations will continue to grow well into the next century. By the year 2015, 91 percent of all the rural people in the world will be living in developing countries. So, if you're interested in rural people and rural development, you're interested in developing countries because that's where most of the rural people will be.

As Bob talked about the changes over time, I also thought about the extraordinary complexity of many of these issues that surround environmental quality, natural resources, agriculture, and technology. It seems to me that we have not yet achieved a way for identifying and taking timely and effective action on environmental, natural resource degradation, and depletion problems. We seem to have difficulty, in many cases, in developing consensus and taking effective action until the problems are very costly or difficult to alleviate. How to speed up the process of moving to appropriate collective action is an important issue. This will involve institutional innovation and advocacy. Perhaps we should look to the NGO institutional mode to encourage more timely advocacy and decision making.

More creativity and innovation is needed to keep natural resource problems from reaching extreme levels where costs for enhancement or reversal are extremely high. The scientific analysis of complex natural resource and environmental conditions is very difficult. These are almost always closed systems, and these systems have feedback loops, they have time lags, and they often have substantial human involvement. They almost always contain complex physical, chemical, and biological processes. These characteristics make it difficult to replicate these systems and to use the traditional scientific method to experiment on them. It's difficult to look at these systems for policy ramifications in a reductionist way. I believe one has to reduce to understand pieces of them, but sometime you have to put these systems together to see what they mean as a whole.

I've been reading about some scientists who have tried to go back in history to understand why certain things happen in natural systems, particularly fish populations. We have learned it's difficult to sort out everything that went on and try to identify what the problems were. At least, multidisciplinary research and analysis is needed. Systems approaches and human capacity building are important elements in making progress in understanding environmental systems. Clearly there is a need for some innovative scientific inquiry. Certainly we can observe these systems, learn from them, and do better than we have. I think we're gaining on that front.

I was recently at a meeting in Washington, D.C., with a number of institutions that work on environmental problems. One of the characteristics that always seemed present in dealing with these problems was conflict. Conflict seems to be involved in these systems because they commonly involve trade-offs among people and among institutions. It seems to me, as we think about working on these issues, that some conflict management skills are needed to better understand and to operate more effectively on environmental systems.

Another set of activities that I believe is important is legal systems, advocacy, rules of the establishment, and legal representation of groups of people. The law, properly understood, administered, and available to people, can be a powerful tool in development and in natural resource and environmental enhancement. We may not have thought about this enough in our development activities, particularly as related to issues of conflict, inequity, and local organization.

My last comments are with respect to some current concerns that I have. Often, people say that the reason we have environmental degradation and environmental and natural resource problems in developing countries is because of the poor people. There may be some truth in that, but I suspect that is one of those slogans on which it would be difficult to base policy until you had a good understanding of certain situations.

The second one that I often hear is that economic growth will resolve the problems—when countries become wealthy enough, they can afford to do environmental and natural resource enhancement activities. While this may be an important part of dealing more effectively with the environment and natural resource activities, I also suspect that there are certain kinds of economic growth that can make environmental problems more difficult.

A third concern is that the answer to these questions in developing countries is aid—that what one needs to do is provide more resources. I believe that we need to think carefully about what form that aid will take to deal with these particular kinds of problems because environmental problems involve activities that many of our institutions have not been heavily involved in over the years such as conflict resolution, legal affairs, local action, and advocacy.

So, we have before us a topic that I believe is very important and one that Winrock has chosen to work on. Let me say a few words about the Winrock program. In 1992, for the first time since Winrock began, the amount of our expenditures in environmental programs equalled expenditures in agricultural programs: We spend about 42 percent of our money in environment and about 42 percent in agriculture. Therefore the intersection between agriculture and environment is an important one. We have a substantial activity in these issues in Winrock. What specifically?

Agriculture's interaction with forests is an important part of Winrock's program for our regional divisions as well as for our large forestry and fuel wood project supported by USAID. We've decided to concentrate our environment and natural resource activities in rural areas and on the natural resource side, so we talk about water. We have a major program in that area supported by the Ford Foundation and other groups. Several of our activities are focused on sustainable agriculture, watersheds, and the kinds of natural resource issues that we see with respect to secondary use of natural resource products.

We plan to keep this emphasis on environment and natural resource problems as we continue to work on environmental policy and training issues and on environmental evaluation and monitoring. We believe that this interaction between agriculture, natural resources, and the environment is increasingly important and one that we at Winrock will be working on for many years in the future.

Discussion

Burton: The floor is open for general discussion.

Hardin: Perhaps it's unfair to ask Bob Havener to comment on Winrock as an institutional innovation, but Bill Dietel has suggested that we need innovations. In the creation of this institution, Bob has been at the helm of what we perceive to be an experiment in institutional development. He has marched it down the road quite a ways. What direction do you see for its future, Bob?

Havener: It's nice to be asked such small questions at important intersections in one's life, isn't it?

Winrock was indeed an institutional innovation, and Lowell Hardin was one of those who helped to conceptualize and structure what it would become. Fortunately one of the things that the founding board members built into the fabric of the organization was the ability to adjust to changing conditions. I believe that when they did so, they had little idea how changed the world would be. They also built into the organization a market responsiveness so that in order to carry out its mission, the institution would have to rely on the beneficence of donors interested in similar issues. Both of those things were extremely important to whatever success Winrock International has enjoyed.

I will remind you that when Winrock was created, agricultural research was a very large item on the donors' agenda. Typical of the projects that IADS was implementing in those days was the one in Bangladesh, where they had 23 expatriate staff members working in a single national program, and the one in Indonesia, where there were a dozen expatriate staff members working in a single national program and where we were training large numbers of young people to come back and assume those important positions.

IADS had earned a reputation in strengthening national agricultural research systems and their linkages with the global networks. Winrock International Livestock Research and Training Center had earned the reputation of being the largest, most expert organization in the NGO community with technical competence focusing on livestock and livestock production systems. And in the eyes of donor community it had a predominant capability on that subject.

Within 2 years, large grants to strengthen national agricultural research organizations and livestock projects were dropped from the agendas of donor agencies, and we had a new organization created with those as the two principal strengths. So, one can imagine the adjustments that we have been given the opportunity to deal with in the last 8 years. But we are meeting the challenge.

We were created to be able (1) to envision and, it was hoped, influence those changes and to bring to bear some judgment on the direction of those changes and (2) to adjust our portfolio and our actions to continue to address emerging development issues. I do think that it is important that we build into organizations the ability to change, to move with market demands, and to adjust to shifts in the development milieu.

This is not without human costs. We too are a changed organization from what we thought we would be 8 or 9 years ago. We've had to recruit people with new skills and knowledge—people who call themselves ecologists and systems analysts—because their skills are necessary to address the problems of today. And we have fewer people who would call themselves livestock specialists and plant breeders, because that is the market in which we operate.

We made a deliberate decision to merge with less endowment capital than we had planned, and we decided that we would try to compensate for that somewhat lower endowment base by working more aggressively for grants and contracts. It has proven difficult to compensate for a lack of endowment by this means. I would counsel people who are thinking about building new institutions and new organizations to be somewhat less parsimonious as they think about the amount of capital, human and financial, that's necessary to assure their future success and to assure the standards of excellence that one wants from such organizations.

New, innovative NGOs normally start small, but size eventually becomes important if they are going to have a significant impact on large development problems. Consequently, as NGOs think about their future, they need to make a fundamental judgment about the organizational size required to wrestle with the problems that they want to address and to build that concept into the institution they create.

I think we've just begun to see the merging of NGOs. We've talked about the proliferation. There are NGOs on virtually every subject that anyone could imagine, and many will not be sustainable. Some are going to pass away because their mandate and mission are no longer relevant or because the money and resource base are no longer sufficient to make them sustainable. Many mergers will occur in the future, and what I would be concerned about are the capital base, the human resource base, and the size and scale of operation that are necessary in order to allow them to have a significant impact on problems of importance.

Goodwin: It has been suggested that there are fashions in thinking about development. You commented on how Winrock has appropriately had to be flexible and responsive to those fashions, and, yet, at the same time it has had to try to lead. What area do you feel is being missed in the current directions that are being taken by the donor agencies? What do you think is being underemphasized?

Havener: First, there is still too little attention being paid to human resource development in most development projects and most places in the world, both in terms of numbers and in terms of kind and character. I think the situation is deteriorating rather than improving.

Because of the very heavy expense of supporting academic degrees, major donors have tended to back away from that aspect of institution and capacity building. This is true even within USAID, and I think that's most unfortunate for American institutions, for the people who are not getting access to the education, and for the laudable goal of building a group of people around the world who are aware of and share some aspects of American values. It is in this way that in the past many became better partners in development.

Second, I fear that there's too little money likely to be invested in the future in maintaining agricultural research competence in our own country and in many countries around the world. I remain concerned that the next 3 billion people will put substantial strain on our production resource base, particularly if we in the developed world are going to place a very high price on the use of that base because of environmental concerns and because we're less willing subsidize our agricultural producers. I think the world could be in a tight food supply situation during the next 10 or 15 years, and I know that if we want to increase the productivity of the existing productive resource base we must continue to make significant investments in agricultural research.

This appears to be a problem at all levels, national and international. Recently, I've been asked to serve on the board of CIAT. Their most difficult problem is budget reductions of 15 percent in each of 3 successive years. That cripples institutions. And it's particularly difficult for research organizations whose time horizon must be long.

Peterson: I would like to answer a little different question and that is: From the point of view of a donor, what do we see that Winrock might be missing? It seems to me that Winrock is a unique institution in the sense that it does marry the people orientation of a NGO operating on the ground with some very strong technical skills and a growing capacity to do policy-type work. But I don't think that Winrock merges those activities as well as it should.

Kellogg: You're right. In part that is because we chose to go ahead with an endowment resource base that was less than it should have been. It takes money to convert projects into programs. As you know, we operate with a USAID-supported project on this and a Ford Foundation-supported project on that and a Rockefeller Foundation-supported project on something else, and these are all important, discrete projects. To integrate them and to maximize the leverage, the knowledge, and the potential output of them takes financial support, among other things. It takes money to buy staff time and talent. It takes money to bring other resources to bear on those individual projects to forge them into a program in an integrated and consolidated way. We also need a clearer vision of how to change our institution to become a technically based NGO. To reinvent Winrock in this way takes us into uncharted territory. It is exciting but uncertain.

Coward: The observation that in Winrock's current activities there is parity between environmental projects and agricultural projects suggests that Winrock is rapidly moving toward being what we might think of as an environment and development institution. What additional modifications will be required in the institutional character or culture of Winrock if in fact it will be an effective environment and development institution?

Kellogg: Some of the modifications that we are looking toward have to do with the human resource base. We are in the process of adding more people with better technical understanding of the environment and development nexus and the institutions that work on that set of problems. In one of our projects in the Philippines, we are evaluating and monitoring a large natural resource conservation program that USAID is supporting. One of our strategies for doing that is to bring a number of short-term people to work on some of these issues in support of a good solid in-country team there. So, we're beginning to reach out to good social, biological, and physical scientists in this field.

Another modification we are looking forward to is to take advantage of the NGOs in the environment-development nexus that have a lot of activity at the grassroots. These grassroots NGOs need support from an organization that has a human resource development perspective and a technical base. So we find ourselves working more and more as an institution that has linkages with NGOs and provides technical assistance, training, and those kinds of services to them.

We also find ourselves being asked to provide more training management and training support within regions as well as from developing countries to the United States. For example, in our office in Manila, we are managing and supporting over 100 Asian students that are going to countries other than the United States to study. Some of them are going to Australia, others to the Philippines, some to Thailand, and others to India.

And then, last, we're becoming much more involved with questions of how to link environmental and natural resource problems with policy development within developing countries. We're beginning work in that area in the Dominican Republic and in Vietnam. Our Environmental Policy and Training project has been doing that in many countries over the past year.

We continue to try to maintain what I believe to be our comparative advantage—a technical base. We don't want to become an advocacy NGO, a grassroots NGO, but a technically based NGO that cares about people, that has wide-range contacts with important decision makers in the public, private, and nonprofit sectors, and that can provide good quality backstopping services for that.

Ruttan: I share Bill Dietel's excitement about what's going on in the NGO community. As one goes around, one can't help but be impressed. But I also have several serious concerns. One is the quality of our knowledge. I've been trying to assemble some literature on NGO programs. The literature is appallingly bad. It's mostly hortatory promotional material. You don't find solid, analytical knowledge. Second, I'm concerned about the emergence of dependent NGOs. As I travel I find, after I get behind the glossy brochures, that there's a great deal of dependency on support from abroad. One of the reasons we've had so much trouble institutionalizing national research systems is that we've taught the science entrepreneurs that head those institutions that it is easier to prospect for resources outside of their country than inside. They never became politically viable at home. When I worked at IRRI, I was impressed by the fact that the Catholic Church in the Philippines, after close to 500 years, was still dependent on clergy from western countries. I think we have to approach this thing thoughtfully if we're going to avoid the emergence of dependent NGOs that are not economically, socially, and politically viable in their own societies.

Dietel: The point you made at the end is the one that maybe interests me the most. And that is the capacity of the outside world to be able to help people in the developing countries establish institutions that can feed those NGOs with funds from within. If you go to a place like Mozambique, there is no great tradition of philanthropy and if you don't want the Mozambiquans dependent upon the Ford Foundation and the Rockefeller Foundation, to say nothing of multilateral aid agencies, then, at the same time that you're tying to help them develop ethically responsible and effective NGOs, you also must help develop their capacity to feed themselves financially to create institutions. I think you can do that and, in fact, it is being done. There are people now putting up what we would call challenge money in Mozambique to create a Mozambiquan foundation. This is long-term stuff, but it will have to happen if the worry that you rightly have is going to be successfully addressed. And these people know little about how to run a nonprofit organization in a way that most of us would find acceptable. But before we become too holy about how well we do it, I remind everyone of the recent scandal with the United Way.

In our country, it is estimated that there are 30 million Americans who serve on the boards of nonprofits. How well they serve, how effective they are as trustees and directors, is a big question. And if we have problems in getting them to be better led and better run, you can imagine what the problems are like outside the

Western world. But that doesn't stop the fact that these people are, in fact, going to create some homebred organizations to do some things that they want to accomplish, which no institution from outside is currently addressing.

Kaul: In Bangladesh, as an example, there is a proliferation of NGOs. Almost 10 a day are being registered, although the government has little capability to screen them. And the aid agencies are tending to funnel more money through NGOs and less through the public-sector agencies. In agriculture, however, the institutional framework that was developed for research has not yet achieved its goals, not even 10 percent of them. But suddenly the aid agencies are shifting not only their funds but also their priorities. With the greater emphasis on NGOs in Bangladesh, in my judgment, the research infrastructure will not be sustained. There is hardly any money even to continue varietal trials. I'm worried that this trend will not be reversible, and there might be tremendous loss to Bangladesh and other countries before we know it.

Berg: It's a testament to Bob Havener's legacy that we are all looking ahead, and one can on this solid and admirable base. In the future, I think some dramatic changes in the concept of sovereignty are coming up. The state is being, let's say, picked upon in two ways: through devolving power in which the NGOs have a certain role and through the evolving power of the multinational system, both the corporate sector and the public sector in the United Nations—the multilateral system. And that has bearing on where pressure and focus should be.

One of the difficulties is that we think of NGOs in the development sense, and a great many Third World NGOs accept our cooperation as development partners. They are really in the empowerment business, and we have to understand both the thrilling opportunities and the risks in that. Nonetheless, the total NGO movement is vibrant and dynamic. It's touching perhaps a tenth, in some areas a fifth, of people in the Third World. It is not an adequate strategy in itself, unless one chooses strategically who one works with, and why, and with better sense than many northern groups have.

The role of advocacy is terribly important. Advocacy, when it comes to putting forth central and important proposals, and how to see problems, and how one should act is extremely important. Expert testimony and explanations of issues to leading citizen groups is terribly important.

I see that not only in gaining an understanding of our national policy, but particularly in the question of where will momentum be generated on world food and hunger issues. And here is the weakness of the United Nations system in comparison with many other issues—Jim Grant's summit for children, Morris Strong's work on environment, and the United Nations conference, Education for All, in Jomtiem—you would not normally think of the top 10 people at FAO as where you would want momentum to be generated on food and hunger issues. It's an institutional and political challenge we can't ignore. How will this great crisis, which is growing, particularly in Africa, be met by the international community? It will take a revamping of that system and a revolution of leadership.

I think that there is a role in public policy advocacy, in the strategic and political sense, from groups like this to lead us out of the wilderness of the international system.

Peterson: One of the pillars that USAID is going to look at is this concept of democratization and empowerment. USAID tends not to be good at grassroots programs. So, we're looking for ways to wholesale our retail grassroots efforts, and there do appear to be problems with working directly through indigenous NGOs. That will create some real difficulties for USAID. It took me 2-1/2 years to get an NGO established in Pakistan because governments, in general, don't like to have indigenous NGOs. And even countries like Bangladesh that have a long history of working with NGOs tend to want to see the money first before it leaves their coffers and goes to the NGO coffers. That then raises the question, what is an NGO? If all the money for an organization comes from the government, it's difficult to really consider it an NGO.

I don't know the solution, but I do think at least for the foreseeable future, people in USAID will be looking for ways of organizing people at the local level and ways of supporting them. That will be much different from the way we've done business in the past.

Grove: I wonder if the problem that Vernon Ruttan was alluding to, and one of the issues that NGOs struggle with, is that they are asked to do things that they're not best qualified to do. Two of the things that donors and others have asked NGOs to do today are human capacity building and research, which are not things they necessarily have great capacity to do. We perhaps have forgotten a lesson from past agricultural development—that we invested hundreds of millions of dollars in U.S. institutions, universities, developmental organizations, research institutions, and through that route developed a capacity to train a cadre of specialists overseas. We've created human capacity in developing countries. When the environment arrived on the scene as a major issue, we decided to take shortcuts. We don't have a cadre of environmental scientists in the United States or any place else that has the developmental understanding of the breadth that the agricultural community does. We've looked at the NGO community perhaps as a short cut for getting there, and that concerns me. I wonder if it is not a high-risk experiment that we're running.

Smith: In most other parts of the world, there are consortiums of NGOs that are working in concert with U.S. groups and that are trying to find those central areas of agreement where they can collaborate, especially on the education of policy makers in multilateral and bilateral institutions. Considering Winrock's four objectives, I can see NGOs also working with Winrock to strengthen their capacity to improve the quality of life for the rural poor. We offer what they don't often have, which is technical expertise at the macro and micro level.

Peterson: The difficulty is that you probably cannot solve the basic environmental issues and some of the basic policy issues without utilizing NGOs. But you will not solve the food equation by going to the NGOs. They do not have the ability to crank out the technology and the technology changes that are necessary to feed the people. And I don't know how to get those two groups together.

James: Bob, you mentioned that two concerns that you have for the future are human resource development and the lack of resources for technology development. What do you think we can do as a community to alert the donor groups that, in fact, this requires urgent action?

Havener: That's a very tough problem. Ambassador Blake, who is a participant in this workshop, has helped us articulate concerns more effectively to organizations like the World Bank, the U.S. Congress, and USAID. He has done a marvelous job—he writes letters to everybody, corresponds, convenes, and communicates. More of us in the NGO community and other organizations will have to follow his lead in aggressively informing the various publics with which we interact.

In our new-found enthusiasm for private-sector involvement, we cannot forget that agricultural research will require investments through the public sector. If we want to focus on regions of the country and on crops and on commodities that are of importance to poor people, we're going to have to continue to support the public institutions, as well as to foster the growth of the private institutions that serve those publics. There's no escaping that challenge in the future.

Kellogg: It seems to me one of the future activities that we need to focus on is national institutional innovations that provide for not only NGOs but agricultural research sustainability in the way of funding.

In India, we were asked by USAID to develop an institutional framework for beginning an India-United States foundation that would make sustained investments in agriculture, technology, research, private business interactions, and so forth. At the same time, the Ford Foundation was investing in India to develop a foundation that would provide ways to sustain activities. I think those are the kinds of useful institutional innovations that are going to be important as we look to sustainability at the national level from now on.

2
Conserving and Enhancing Soil Resources for the Future

Introduction

THURMAN GROVE
North Carolina State University

I see three challenges for future soil researchers and managers and perhaps the current practitioners. First, how do we keep soils in place or should we even worry about keeping them in place? Each year about a billion tons of soil are carried to the sea by the rivers of the United States. Some claim that for every kilogram of maize we produce in the Midwest, we lose 2 kilograms of soil to wind and water erosion. FAO estimates that we lost 203 million hectares of arable, rainfed land in Africa between the 1975 and 2000, principally to wind and water erosion. We certainly have a lot of knowledge and technology for conserving soil, but for some reason it doesn't seem to work, and I think we need to pay more attention to why it doesn't.

The second issue is, how do we manage soils as renewable natural resources? We're still in the chemical stage of agriculture. The use of chemicals has produced some astounding results. But our use of soil amendments is inefficient and often has environmental side effects that we don't like. Perhaps it's time to recognize that soils are ecosystems complete with biological communities that are the principal source of their renewal capacity. We need to learn more about how to manage these biological communities, rather than being their surrogates.

A third issue is that we haven't been very effective at developing technologies for sustainable use of soils that take into account the economic and social aspirations of the users. Rather, our technologies are largely driven by the professional biases of the discipline of soil science.

All of these challenges indicate that the field of soil science needs to broaden its views of soils research and development and integrate them more fully in a broader suite of developmental and productive activities.

Conditions for Achieving a Sustainable Global Agricultural System

PIERRE CROSSON
Resources for the Future

This paper is addressed to the broad question: Can the global agricultural system satisfy rising demand for food and fiber to the year 2030 at "acceptable" economic and environmental costs? Costs are understood to include those borne by people living between now and 2030, as well as costs imposed on subsequent generations by management of the system between the 1990s and 2030. The paper thus is about the sustainability of the global agricultural system in the sense that one's judgment of whether the system is or is not sustainable likely would depend on whether the answer to the question about acceptable costs is yes or no. I chose 2030 as the target year because it is far enough into the future to expect sustainability issues to arise, but not so far as to make speculation fruitless because of the high uncertainties involved.

The emphasis on economic and environmental costs in the question addressed indicates that the paper is focused on the supply side of the global agricultural system. However, the economic and environmental costs of increasing supply depend importantly on the rate of increase in global demand for food and fiber. Other things being the same, holding economic and environmental costs within acceptable limits is more difficult at higher rates of demand growth than at lower rates. More specifically, the pressure of agriculture on the natural resource base and environment is greater with higher rates of global demand growth than with lower rates. As a result, the capacity of the agricultural system to respond satisfactorily to rising demand—the sustainability of the system—cannot be assessed without some notion of the likely growth in global demand for food and fiber.

The supply response to increasing demand, and the consequent behavior of economic and environmental costs, will depend on the quantity and quality of the global resources that can be mobilized for agricultural production. I concentrate here on five categories of resources: land, water, plant genetic resources for crop breeding, climate resources, and knowledge about agricultural production embedded in people, institutions, and technology. At any given time, these resources may be combined in relationships of both complementarity and substitutability, depending on technical and economic conditions. Over the past 50 years, however, knowledge has increasingly substituted for the other resources because it has proved far easier to increase the supply of knowledge than the supply of other resources.

All present evidence, including that marshalled for this report, indicates that this trend in the relationship between knowledge and the other resources will continue into the indefinite future. Indeed, the critical question with respect to agricultural sustainability is, what are the long-term limits, if any, of substitution of knowledge for the other resources employed in agricultural production? This view of the sustainability issue determines the structure of the paper.

The Demand Scenario

To assess the sustainability of the global agricultural system, we need some notion of the demand pressure that might be brought on the system. We need a demand scenario. The scenario sketched here deals primarily with future global demand for grains. At present, grains are estimated to account for roughly one-half of global consumption of food energy; and, as animal feed, grains also account for a rising share of global protein consumption. Although soybean is an important high-protein food for people and, particularly, animals, it is a special case. Because its production is concentrated in four countries (the United States, Brazil, Argentina, and China), it is not considered in the supply-side perspective taken here. The demand for animal products is assumed to be reflected primarily in the demand for feedgrains. However, the resource and environmental impacts of forest clearing to grow crops or graze animals are considered in the supply-side assessment.

Projections of the growth of food demand over several decades typically are based on projections of population growth and per capita income. That is the procedure employed here. United Nations and World Bank projections of population coincide in showing that from 1990 to 2030, some 90 percent of global population growth will be in developing countries. Moreover, only in those countries do increases in per capita income add much to demand for food measured at the farm gate. Most people in the more-developed countries are sufficiently well fed that additions to per capita income induce little if any increase in food demand.

The combination of these population and per capita income factors is such that, in the demand scenario sketched here, over 90 percent of the growth in food demand occurs in the developing countries (Table 1). Global demand doubles from the late 1980s to 2030. In the developing countries, demand grows 2.7 times. But their rates of growth in demand for wheat and rice slow markedly from 1988/89 to 2030 compared with the rates that prevailed from 1979/81 to 1988/89. The demand growth rate for coarse grain, however, almost doubles (from 1.7% to 3.2%). This change in the pattern of demand for the three types of grain is comparable to the experience in South Korea and

TABLE 1
Annual grain consumption in the less-developed and more-developed countries, 1979/81-2030.

	Quantity (million t)			Annual increase (%)	
	1979/81	1988/89	2030	1979/81-88/89	1988/89-2030
		Less-developed countries			
Wheat	195.6	265.6	770	3.7	2.3
Rice	249.4	309.2	634	2.6	1.3
Coarse grains	260.8	299.7	946	1.7	3.2
Total	705.8	874.5	2,350	2.6	2.3
		More-developed countries			
Total	--	802.5	947	--	0.4

Source: Crosson and Anderson (1992).

other parts of East Asia over the last couple of decades (Mitchell 1991). The region generally enjoyed rapid urbanization and fast-rising per capita income, especially in South Korea, accompanied by a rapid increase in demand for wheat, and subsequently for coarse grain, in relation to demand for rice. Mitchell (1991) argues that a shift to a more urban, higher income society induces increased demand for a higher quality diet, particularly with respect to animal protein, and the observed shift in patterns of demand for grain reflects this.

In the rest of this paper, the demand scenario sketched above is taken as the challenge confronting the global agricultural system over the next 40 years. Can the system increase the supply of food in response to the demand scenario at economic and environmental costs consistent with the aspirations of people all around the world, especially in the developing countries, for a rising standard of welfare? That is the broad question addressed. Because almost all of the projected increase in grain consumption is in the developing countries, the focus is mainly on them.

The Supply Response

The global agricultural system must somehow mobilize the social capital necessary to satisfy the demand scenario at economic and environmental costs consistent with sustainability. The critical components of the social capital are natural resources and human-made resources. Natural resources are land, water, plant and animal genetic resources, and climate resources. Human-made resources are the knowledge embodied in people, technology, and institutions.

Because knowledge, to some critically important but poorly understood extent, is a substitute for natural resources, I proceed in two steps. I first consider whether the supply of natural resources, *with no increase in the supply of knowledge resources*, could be increased enough to accommodate the demand scenario at economic and environmental costs consistent with sustainability. The answer is a definite no. I then proceed to consider the conditions necessary to increase the supply of knowledge enough to sustainably satisfy the demand scenario. The conclusion can be stated in two parts: (a) the opportunities for expanding knowledge are likely to be sufficient to achieve sustainability *if* we have the wisdom and political will to commit the resources needed to generate the new knowledge; (b) if we fail to make that commitment, sustainability will not be achieved, because the potential supply of natural resources is far too limited. The key to sustainability is knowledge.

The Supply of Natural Resources Given Present Knowledge

Among the four natural resources considered—land, water, plant and animal resources, and climate resources—most attention here is given to land. In my judgment, the demand scenario will put the heaviest pressure on the supply of land, and the land supply response may be more problematic than the response of the other natural resources.

The land resource

Land supply has both a quantitative dimension and a qualitative dimension. With respect to quantity, FAO estimates that in the 1980s there were about 1.5 billion hectares of land in crops with a roughly equal amount in forest, pasture and range with some potential for conversion to crop production.

As a first approximation, the estimate of potential cropland suggests that most of the increase in the global demand scenario could be met by bringing the potential land into crop production. The first approximation, however, is misleading. Surely the amount of land now in forest, pasture, and range that could be converted to crop production at acceptable economic and environmental costs is far less than the FAO estimates indicate. One reason is the geographic distribution of the potential cropland. The land is quite unevenly distributed among developing countries, 45 percent of it being in Africa and 49 percent in South America. In principle, the better endowed countries of Africa and South America could export food to land-short Asian countries. However, for domestic political and national security reasons, Asian countries are not likely to regard a hectare of uncultivated potential cropland in Africa or South America as equivalent to a hectare within their own borders.

An additional constraint on the supply of potential cropland is that most of such land in South America is in the humid tropics, and most of it is now in forest. Compared with currently cultivated land, the potential land is far from domestic and foreign markets, and road, rail, and air connections to those markets are poor. In Africa, because of a poorly developed road and rail system, this lack of transport infrastructure is perhaps even more of a constraint to opening new land to crop production than in South America.

Much of the potential cropland in Africa and South America could be exploited only by clearing land now in forest. This would involve some opportunity cost measured by the value of forestry services that conversion would foreclose. Conversion of range to crop production also would exact an economic opportunity cost in terms of lost animal production and the complementarity that this brings to the cropping phase of rotations. It is clear that the estimates of potential cropland do not take these opportunity costs into account.

Tropical forest clearing likely would also incur environmental costs from losses of biological diversity and from additions of carbon to the atmosphere, which contribute to global warming. These costs generally are not reflected in markets, so no one knows their present or potential future magnitude. Nonetheless, the costs are believed to be high by influential members of the world community, and widely publicized efforts are under way to persuade the Brazilian and other governments in tropical areas to slow, if not halt, tropical forest clearing. So far these efforts have met with little success, but there is every reason to believe they will continue. The pressure to control deforestation will mount and will likely have more effect. In this case, the realizable cropland potential of South America, and possibly Africa, would be considerably less than the FAO numbers suggest, other things being equal.

Finally, the conversion of present and potential cropland to urban and other built-up uses will constrain the future quantity of croplands. The scientific literature reveals little about the amount of land in urban and other built-up uses on either regional or global scales. Consequently, projections of the amount of potential cropland that might be converted to urban uses over the next 40 years are subject to wide margins of error. Working with the skimpy data available, and with the help of certain heroic assumptions, Crosson and Anderson (1992) concluded that by 2025 urbanization might take about 5 percent of present and potential cropland in the developing countries.

Considerations of land quality likely would also limit the land supply response. There are two aspects to the land quality constraint. One is that the potential cropland is inherently less productive in crop production than land already in that use. This should not be surprising. The better soils would naturally be brought into crop production first. It follows that a hectare of potential cropland is not the equivalent in productivity of a hectare of land already in crop production.

Land degradation—the other aspect of land quality—is more important than the quality difference between present and potential cropland because it affects both land already in production as well as potential cropland. Moreover, there is more disagreement about the extent of land degradation. That disagreement has much significance for policies to achieve a sustainable agricultural system. For these reasons, I treat the land degradation issue at some length.

A review of the literature on land degradation reveals both a deep concern about the consequences of degradation for the sustainability of the global agricultural system and a recognition of how little is reliably known about present rates of degradation and their long-term implications for sustainability. The concern dates at least from the United Nations Desertification Conference in Nairobi in 1977. In a follow-up study designed to estimate changes 5 years after the conference, Mabbutt (1984) concluded that by 1982 land degradation (called desertification in the paper) annually reduced the productivity of some 20 million hectares to zero, that another 6 million hectares annually were converted to "wasteland," and that the number of people living in "desertified" areas increased 35 percent from 1977 to 1982.

Nelson (1988), in a study for the World Bank, reviewed the evidence on land degradation, including that presented in Mabbutt's 1984 paper, and found it often ambiguous and contradictory. Some studies about the advance of deserts in Africa and India were contradicted by subsequent studies that showed advances in periods of drought followed by retreat when the rains returned. Nelson (1988, 1) concluded:

- The extent and severity of land degradation are not as well known as commonly believed. The evidence on this is "extraordinarily skimpy."
- The degree of professional agreement about the extent and causes of land degradation, and about solutions to it, is generally overestimated.
- The irreversibility of land degradation processes probably is overestimated, although serious losses have occurred in some areas.
- The image of land degradation too often is one of advancing sands instead of, more properly, one of "pulsating deteriorations," sometimes with reversals or at least long-term remissions.

Nelson's paper illustrates the theme that little is known about the extent and severity of land degradation. Dregne (1988, 679) notes that estimates of land degradation, including his own, are based on "... little data and much informed opinion." Writing specifically of soil erosion and its productivity effects, Dregne (1988, 680) asserts that "there is an abysmal lack of knowledge of where water and wind erosion have adversely affected crop yields." He notes that the equations for calculating wind and water erosion were developed for use in the temperate zone (more specifically, the U.S. Midwest), and their accuracy for the tropics is uncertain.

El-Swaify, Dangler, and Armstrong (1982, 1), authors of the most comprehensive published study of soil erosion in developing countries, assert that "... there is little or no documentation of the extent, impact, or causes of erosion ..." in tropical environments. Lal and Okigbo (1990) share these several views about the lack of reliable land degradation data. They note, with respect to Nigeria, that most of the evidence on soil degradation is from analysis of soil sampled from experiments at research stations. "There is little, if any, research done on farmers' fields to provide concrete and reliable quantitative information on the rate, extent, and distribution of soil degradation" (Lal and Okigbo 1990, 7).[1]

---

[1] Oldeman, Hakkeling, and Sombroek (1991) prepared a map, with an accompanying explanatory text, showing the state of human-induced degradation of the world's soils. The total degraded area is assessed as 1,964 million hectares, 15% of the total mapped area of 13,013 million hectares. Water erosion accounts for 56% of the 1,964 million degraded hectares, wind erosion for 28%, chemical degradation for 12%, and physical degradation for 4%. For reasons given in Crosson and Anderson (1992, 33-34), the significance of these estimates for the global supply of cropland is not clear. However, more research to advance the initiative started by Oldeman, Hakkeling, and Sombroek should have great value in adding to knowledge in this important but poorly understood area.

The research on land degradation indicates that the most important sources of degradation, by far, are erosion by wind and water. Stocking (1984), after systematically reviewing this research, concluded that almost all of it had been done in the United States. Stocking's conclusion still is valid.

Before the late 1970s, all the research in the United States on erosion-productivity relationships was conducted on small experimental plots. The small scale of these experiments and the highly specific soil and climatic conditions under which they were undertaken made it impossible to extrapolate the results to national-scale estimates. Surveys undertaken by the U.S. Soil Conservation Service in 1977 and 1982 provided the data needed to make such estimates. Three models were developed to do this, the productivity index (PI) model (Larson, Pierce, and Dowdy 1983; Pierce et al. 1984), the erosion productivity impact calculator (EPIC) (Alt, Osborn, and Colaccio 1989), and a third model developed at Resources for the Future (Crosson 1986). PI and EPIC simulate the responses of crop growth to changes in soil characteristics—pH, water-holding capacity, nutrient supply, bulk density, etc.—under the impact of soil erosion. The RFF model is a regression type that accounts for intercounty differences in crop yields among a sample of counties in the Midwest as a function of intercounty differences in soil erosion, soil depth, fertilizer application, and other inputs.

Working with EPIC, Alt, Osborn, and Colaccio (1989) estimated that, if cropland erosion in the United States were to continue at 1982 rates for 100 years, national average yields at the end of the period would be about 3 percent less than they would be in the absence of erosion. Pierce et al. (1984) used the PI model to estimate the effect on yields of maize on 40 million hectares of cropland in the Midwest under the same erosion conditions and got a 4 percent yield decline at the end of 100 years. Crosson (1986) used the Resources for the Future model to estimate the effect of 100 years of 1982 erosion on maize, wheat, and soybean yields in the Midwest. He found that maize and soybean yields would be 5 percent and 10 percent less, respectively, after 100 years than they would be in the absence of erosion. Wheat yields would be unaffected.

Two features of this research stand out. One is the close similarity of the results, particularly in view of the quite different modeling approaches taken. (EPIC is a far more complex model than PI). The second notable feature is the smallness of the erosion effect on soil productivity, even after 100 years at rates that the U.S. Soil Conservation Service officially regards as too high on close to half of the nation's cropland.

The finding that soil erosion in the United States almost surely is not a long-term threat to the supply of agricultural land does not mean that the erosion threat is similarly small in the developing countries. Authorities on characteristics of the tropical soils found extensively in the developing countries agree that those soils generally are more susceptible to erosion than temperate-zone soils and that a given amount of erosion on tropical soils is likely to induce a greater decline in soil productivity (Stocking 1984; Lal 1984). Consequently I do not suggest that because erosion is not a serious productivity threat in the United States it is not also in developing countries. Unfortunately, the present ignorance about soil erosion and its productivity effects in the developing countries precludes firm judgments about this important issue.

In summary, in developing countries the amount of land now in forest, pasture, and range that could be converted to crop production at acceptable economic and environmental costs, within the existing knowledge regime, is almost surely far less than the 2.7-fold increase in crop demand in the demand scenario.

The water resource: potential for increased irrigation

In 1986, the global area of irrigated land was 253 million hectares, 2.5 times more than in 1950 (World Bank/UNDP 1990). The 253 million hectares were 17 percent of global cropland, but accounted for more than one-third of total world food production. In the less-developed world, almost 60 percent of rice and 40 percent of wheat—by far the major crops—came from irrigated land (World Bank/UNDP 1990, 3). FAO estimates that, from the mid-1960s to the mid-1980s, the expansion of irrigation accounted for over one-half the increase in global food production (cited by World Bank/UNDP 1990, 103).

On a global scale, the amount of potentially irrigable land in the late 1980s was estimated to be about 50 percent of then-irrigated land (World Bank/UNDP 1990). The potential expansion in the developing countries was 59 percent. For a variety of reasons, these estimates are likely to greatly overstate the amount of newly irrigated land that can be brought into production at acceptable economic and environmental costs.

The estimates of potential reflect judgments of how much additional land has the soil, climatic, and terrain features suitable for irrigation. The economic and environmental costs of actually irrigating the land, however, are not taken into account. World Bank studies (e.g., World Bank/UNDP 1990) indicate that these costs could be high. One of the reasons for expecting high economic costs is the chronically inefficient management of large-scale public irrigation systems all around the developing world. Another such reason is that the lowest cost irrigation sites already have been developed.

Increased demand for irrigation in the developing countries also is likely to confront stronger competition for water to protect and enhance environmental values. The combination of increasing population, particularly urban population, and per capita income in developing countries seems sure to stimulate rising demand for water-based environmental services. By the second or third decade of the next century, this demand could be a powerful competitor with agriculture for water, thus constraining the further expansion of irrigation. For a more detailed assessment of these constraints to further growth of irrigation see Crosson and Anderson (1992).

The genetic resource

Crops and animals are under continuing assault from a host of pests and diseases and suffer from the vicissitudes of often harsh climates. Maintenance of present levels of crop and animal production requires a sustained effort by plant and animal breeders to develop new strains better able to resist relentless attacks. Expanding production on the scale indicated in the demand scenario will require an even more intensive effort by breeders.

To be successful in this challenging undertaking, breeders must have access to a broad range of genetic material on which they can draw to develop more resistant and more productive plants and animals. The plant and animal gene pool, therefore, is a critical resource for achievement of sustainable agricultural production. The literature on the role of genetic resources in agricultural development deals mainly with plants. That is the focus here, too.

Hawkes (1985) assessed the status of the international system for managing plant genetic resources and found it generally good. The focus of his assessment was the member institutions of the Consultative Group on International Agricultural Research (CGIAR) and on the 13 food crops for which they were at the time responsible. These crops include all those of importance at present in the world's diet. Hawkes had a number of suggestions for improving the systems, but he concluded that, by and large, it was ". . . proceeding along the right lines" and that, apart from some mild criticisms, ". . . the general record of the CGIAR in genetic resources work is very impressive indeed." (Hawkes 1985, 101-102)

McNeely et al. (1990) take a somewhat less sanguine view than Hawkes of the present status of the plant genetic resource and the system for maintaining it. Citing Plucknett et al. (1987), they note that, for many of the major world food crops such as wheat, maize, oats, and potatoes, more than 90 percent of the genetic variation in land races is now protected in seed banks and that most of the work needed to do this for other species, such as rice, sorghum, and millet, would have been done by 1990. However, citing Peeters and Williams (1984), McNeely et al. (1990) point out that, of the 2 million accessions of plant genetic material in seed banks around the world, an estimated 65 percent lack basic data on source, 80 percent lack data on such characteristics as methods of propagation, "and 95 percent lack any evaluation data such as responses to germinability tests. Extensive data are held on only 1 percent of the specimens, and it is feared that a substantial proportion of the accessions not tested for germinability might be dead" (McNeely et al. 1990, 65).

The plant quarantine issue. It is not always sufficiently appreciated how dependent the world food system is on the international exchange of plant genetic material. Over 98 percent of agricultural production in the United States, for example, is based on genetic material native to other areas. In the Americas as a whole, half of total crop production is derived from genetic material native to Asia or Africa. Similarly, African crop production is 70 percent derived from Asia or the Americas, and Asian production 30 percent from American or African sources. (McNeely et al. 1990, 57)

The system for maintaining plant genetic resources described by Hawkes (1985) recognizes the interdependence among regions for these resources. The transfer of plant materials from one country to another inevitably carries the risk that plant diseases and pests will be spread to the receiving country. Virtually every country has procedures to inspect and, if necessary, quarantine plant material to protect the nation against this risk. There is in this a potential conflict between plant breeders and quarantine managers, which has implications for the use of genetic resources to expand agricultural production.

Plucknett and Smith (1988) address this set of issues. They cite much evidence that the international transfer of plant genetic material has, in fact, resulted in the spread of viruses that attack various food crops. Despite this, Plucknett and Smith judge that the quarantine limitations placed on international trading in plant genetic material are not a serious threat to continued breeding of improved crop varieties.

Conclusion about the crop genetic resource. The work of Hawkes and of Plucknett and Smith suggests an optimistic view of the present status and future prospects of plant genetic resources. Although McNeely et al. (1990) are more critical of the system for managing plant genetic resources, their assessment is not fundamentally different from that of Hawkes.

There is, of course, no guarantee that the system will stay indefinitely in good health. Should funding for the CGIAR system decline, for instance, the vitality of the seedbank system likely would decline with it. But if funding remains adequate and basic research on plant genetic behavior also is supported, then it seems that plant genetic resources should not seriously constrain the development of the more-productive crop varieties that will be needed to respond adequately to the demand scenario. However, because the resource already is reasonably well managed, improvements in management would do little to expand the supply of the resource.

The climate resource

Characteristics of the climate, especially the average amount and seasonality of precipitation and temperature, directly affect the quantity of agricultural production. The climate is a resource to agriculture, therefore, in the same sense that land, fertilizer, and labor are.

There is now a strong scientific consensus that the global and regional climates are likely to change over the next 40 to 50 years because of the enhanced warming effect of so-called greenhouse gases—carbon dioxide, methane, nitrous oxide, and chlorofluorocarbons. The combination of these gases, equivalent to a doubling of atmospheric carbon dioxide, is expected to increase global average temperatures some 2° to 4° C from present levels by roughly the middle third of the next century (Parry 1990).[2]

What might global and regional climate change mean for agricultural production? Because of the great uncertainty about regional climate change, no clear answer is possible. The studies that have been done, however, suggest that prospective climate changes likely will reduce the supply of climate resources in some regions, but increase it in others. Where supply decreases, farm-level adjustments should moderate the negative impact. The important point, however, is that the supply of climate resources is not likely to add anything to the capacity of the global agricultural system to meet the challenge of the demand scenario.

Economic and Environmental Costs

Over the next 40 years, how would seeking to meet the demand scenario within the existing knowledge regime affect economic and environmental costs of global agricultural production? Another way of phrasing the question

---

[2] The literature on climate change has exploded in the past 5 years, and no effort was made to cover it in preparing this paper. Rather, I rely on the earlier distillations of this literature by Martin Parry (1990). He was lead author of the assessment of these consequences by the Intergovernmental Panel on Climate Change, and his cited 1990 book is based on that assessment.

is this: Holding the supply of knowledge resources fixed, what would be the economic and environmental costs of meeting the demand scenario by increasing supplies of land, water, and genetic resources, assuming that the supply of climate resources, on a global scale, is unchanged?

Economic costs

No attempt is made to estimate economic costs quantitatively, a task quite beyond the scope of this paper, even if the data and models for estimating such costs were available, which they are not. Work done by Binswanger (1989) and others is suggestive, however, of the possible impact of the demand scenario on economic costs if the supply response of farmers was limited to bringing more land, water, and other resources into production within the existing knowledge regime. In brief, the studies suggest that under these supply circumstances, farm-level prices would have to rise 500 to 1,000 percent to induce the necessary supply response to the demand scenario in the developing countries. The world community would surely consider such economic cost increases to be inconsistent with any reasonable concept of sustainability.

Environmental costs

Environmental costs of agricultural production are unpriced and are paid by someone other than the farmer who generates them. The lack of prices for environmental costs creates a major problem in estimating the importance of the costs. The problem is well illustrated by pesticides. Pesticides are designed to kill things: weeds, insects, diseases, and fungi. Over the past four decades, they have done this well and, in the process, made an enormous contribution not only to increased world food production but also to human health, e.g., through control of the malaria-carrying mosquito. But pesticides often affect unintended as well as intended targets and can, therefore, be a threat to human health, wildlife, and whole ecological systems. It is these unintended damages that constitute the environmental costs of pesticides. But nowhere are there comprehensive estimates of the monetary value of the costs that provide a basis for judging their importance relative to other costs of agricultural production or their trend over time.

Although the present levels of, and trends in, environmental costs of pesticides are not known for any country, it seems reasonably clear that, with the supply of knowledge resources fixed, the demand scenario for 2030 would induce farmers, particularly in developing countries, to increase the application of pesticides, both in total and per hectare. A doubling of global food demand

and a 2.7-times increase in developing countries would seem certain to stimulate this response, in the absence of new knowledge about pest management. In this event, the probability of rising environmental costs of pesticides over the next 40 years would appear high.

Sediment carried in runoff from farmers' fields may also impose significant environmental costs. Eckholm (1976) published evidence of substantial damages from sedimentation in countries around the world, but especially in developing countries. He noted increased flooding in the Indus plain of Pakistan, in parts of India, Thailand, the Philippines, Indonesia, and Malaysia, and in the Cauca Valley of Colombia, and he attributed it to rising streambed levels because of sediment from eroded land upstream. Eckholm also noted that accelerated siltation of the Mangla reservoir in Pakistan has reduced the expected life of the facility by one-third to one-half and that reservoirs in other Asian countries also were threatened.

Eckholm's evidence was anecdotal and did not provide the basis for a comprehensive estimate of the costs of sediment damage in either the more developed countries or the developing countries. Subsequent studies have not filled this knowledge gap, although they appear to support Eckholm's argument that the costs of sediment damage are high, particularly in developing countries. The World Bank (1991a), for example, cited a study of eight reservoirs in India that found that rates of siltation are substantially higher than design rates and that on average actual reservoir life will be only 35 percent of the design life unless siltation rates are reduced. Similar accounts of sediment damage are reported in other studies (e.g., World Bank 1992, with respect to China).

As in the case of pesticides, it appears that farmers' responses to the demand scenario would increase sedimentation costs. With the supply of knowledge resources fixed, farmers inevitably would respond to the growth of demand by converting forestland and rangeland to crops. Historically this has increased erosion and sedimentation. Although more sedimentation does not necessarily imply higher environmental costs, the scale of the response to the demand scenario almost surely would force a rise in costs.

Another environmental cost is the loss of biological diversity. While the system for protecting the plant genetic resources needed to develop new crop varieties works reasonably well, those genetic resources are only a small part of the plant and animal gene pool of present and possible future value to human beings. Wilson (1989, 108) refers to this pool as ". . . a potential source for immense untapped material wealth in the form of food, medicine, and other commercially important substances."

Without attempting quantitative estimates of losses of biological diversity, McNeely et al. (1990, 41) cite Wilson (1988) as the authority for an assessment that "... by many indications, the world is already experiencing extinction rates of greater scale and impact than at any previous time in the earth's history."

The poor quality of the data on tropical deforestation and on whatever losses of biological diversity this might entail does not permit confident judgments about trends in the environmental costs of diversity losses. With respect to the future, McNeely et al. (1990) believe that rising world population and economic activity are likely to increase the rate of deforestation. I believe this is the implication also of the demand scenario. In response to that scenario, given no increase in knowledge resources, farmers in tropical areas likely will have incentive to continue to clear land for crop and livestock production. One consequence would be rising costs of losses in biological diversity.

Conclusion on economic and environmental costs

The discussion of the two kinds of costs points unequivocally to the conclusion that, with the supply of knowledge resources fixed, the demand scenario would imply rising economic and environmental costs of agricultural production in the developing countries and probably in all countries with an active agricultural sector. Globally and in most countries, the supplies of land and water could not be increased enough to accommodate the demand scenario at acceptable economic costs, and the drive to increase production inevitably would result in increased use of agricultural chemicals and greater land clearing that would force equally unacceptable increases in environmental costs.

The Knowledge Resource

The inadequacy of present and potential supplies of natural resources for agriculture implies that if the demand scenario is to be met at acceptable economic and environmental costs, it will only be because the supply of knowledge is increased enough to do the job. Institutional improvements, such as establishment of clear and enforceable property rights in land and water resources, could increase the productivity of land and water resources somewhat, but they would not basically change the prospect for rising economic and environmental costs, given the demand scenario.

This argument for investing in knowledge rests on two propositions:

- that the elasticity of the supply of knowledge relevant to agricultural production is high relative to the elasticity of supply of natural resources
- that across a wide range of situations, the elasticity of substitution of knowledge for natural resources is high

No effort is made here to demonstrate the validity of the two propositions, beyond noting that they are consistent with the experience of global agriculture over the past 40 or 50 years. That experience was characterized by an unprecedented increase in global agricultural output accompanied by a massive expansion of knowledge embodied in people, technology, and institutions and, by comparison, a modest expansion in the quantities of land and water devoted to agricultural production.

The discussion in this section is focused on the conditions necessary to expand knowledge pertinent to agricultural production on the requisite scale, at the requisite time, and of the requisite sort. Most attention is given to increasing knowledge embedded in technology and management practices, not because human capital and institutions are less important, but because the processes for directly developing new technology and management practices are in the front line of the knowledge battle. They depend, of course, on institutional effectiveness and, especially in the longer term, on human capital of the requisite quality. The roles of both education and institution-building in development are crucial.

Importance of yield-increasing technology

Rice. Rice is predominantly an Asian phenomenon, and is expected to remain so into the foreseeable future. Asia currently accounts for roughly 90 percent of global production and consumption of rice, and the World Bank's International Economics Department (1990) expects this to hold also into the next century. Future increases in rice production, therefore, will depend overwhelmingly on increasing production in Asia.

Because the land constraint will remain particularly binding in Asia, I assume that the entire growth of demand for rice will have to be met by increasing yields. How much of an increase is that? In the demand scenario, demand in developing countries for rice rises an average of 1.7 percent annually from 1988/89 to 2030. The rate of increase in the first 15 years is about twice the rate over the rest of the period. The question for rice technology therefore is: Can an average annual yield increase of 1.7 percent be achieved over the next 40 years at economic and environmental costs that Asian societies would find acceptable? Experience in the 1980s suggests a cautiously optimistic answer to the question. Global rice yields increased 2.4 percent annually from

1978/82 to 1985/89 and production rose 11 percent (Foreign Agricultural Service 1990). FAO data indicate that consumption of nitrogen fertilizers in Asia increased from 19.5 million tons in 1979/80 to 33.4 million tons in 1988/89. Much of the increase was surely applied to rice land. Use of pesticides in Asia also increased sharply in the 1980s, and much of this also must have been applied to rice. No firm conclusion is possible about the environmental costs of these increases in intensity of use of fertilizer and pesticides in Asia, except that the costs doubtless rose.

The yield and economic cost experiences of Asian rice production in the 1980s support a tentative judgment that the region can increase rice yields and production over the next several decades at the requisite rates and on socially acceptable terms, bearing in mind the uncertainty about recent and possible future environmental costs. This judgment needs further consideration, however, because there now is widespread concern among rice specialists that the recent generally favorable yield experience may not be sustainable. The concern is based on evidence of growing environmental difficulties and stagnant or even declining yields in some areas of Asia (Pingali 1991) and, more basically, on the fact that the yield *potential* of rice varieties has not increased since the 1960s, when the International Rice Research Institute released the first modern varieties for general use (Ruttan 1991, Pingali 1991). The rapid increases in rice yield since the 1960s were achieved by exploiting that initial potential, which, in turn, came mainly through increasing the rice harvest index, i.e., the ratio of grain to aboveground plant biomass.

With allowance for these concerns, Anderson and Herdt (1989) are cautiously optimistic that if rice research receives funding not less than at present, average yields can continue to increase relative to potential yield and that over the longer term some increase in potential also can be achieved.

Meeting the yield challenge to rice will require that Asian agricultural research institutions, including IRRI, have a clear and broad perspective on the longer term problems facing the Asian rice economy and that they receive financial support from the international donor community and from Asian governments sufficient to do the job. Recent experience in this regard is mixed and is anything but uniform across Asia (Pardey, Roseboom, and Anderson 1991, 226-43, 416). It must be observed, however, that most Asian governments have done better than governments in other regions in protecting and sustaining their research infrastructure. Because rice is overwhelmingly an Asian crop, this support for research is grounds for encouragement that the future challenge can be met.

Wheat and coarse grains. Unlike rice, production and consumption of wheat and coarse grains are scattered around the world. Although some regional shifts in production of these crops are likely over the next several decades, especially when account is taken of possible changes in regional climates and of the probability of agricultural reforms in the Commonwealth of Independent States, production and consumption of the crops are not likely to be less geographically dispersed than they are now.

Over the past couple of decades, technological advance in developing countries' wheat production, as measured by the rate of yield increase, has been remarkable, no doubt reflecting in part the fact that the modern wheat varieties are particularly well suited to irrigated production, which constitutes some 42 percent of the wheat area in developing countries (CIMMYT 1989, 2). The achievement in yield performance suggests a well-developed capacity to develop higher yielding varieties of wheat.

Taking account of potential cropland in Latin America and Africa, Crosson and Anderson (1992) concluded that an average annual increase of 1.9 percent in wheat yields over the next 40 years would be sufficient to meet the demand scenario for wheat and that this should be quite achievable. Implicit in this relatively optimistic long-term view is a judgment that there will be considerable contributions from biotechnology-based innovations and through improved crop management as farmers become more skilled in realizing the potential that already exists for yield improvements in wheat. Needless to say, all such advances depend crucially on the sustained and strong support for investments in national and international research and extension systems.

One reason for giving explicit attention here to coarse grains is that the demand scenario indicates a faster rate of increase in global demand for coarse grains over the period to 2030 than for wheat. This pattern is based heavily on Asian experience showing that, with rising per capita income and urbanization, demand for coarse grains as animal feed grows relative to demand for wheat and rice. The demand scenario assumes that this pattern will be increasingly evident not only in Asia but in Latin America and Africa as well.

The faster expected growth in demand for coarse grains in developing countries and the so far slower rate of advance in coarse-grain technology suggest that, in considering prospects for developing higher yielding technologies, major attention must be given to coarse grains. Most coarse grain production in developing countries is under rainfed conditions, although comprehensive data are lacking. Yields of coarse grains generally have not increased as rapidly as wheat yields in the developing countries over the past few decades. The reasons for this are not obvious, but Anderson, Herdt, and Scobie (1988,

24), focusing on maize, by far the dominant coarse grain, offered some explanations. One is that maize is grown under highly diverse conditions, so that individual varieties have to be adapted to narrow circumstances. By contrast, lowland rice and irrigated wheat are grown under much more homogeneous natural conditions. A related reason is that maize is grown throughout the less-developed world unlike rice, which is mostly an Asian crop, and most wheat, which, among the developing countries, is grown mainly in India, West Asia, the Middle East, and North Africa. A consequence of the greater geographical dispersion of maize production is that international maize researchers have had to establish and maintain effective working relationships with a large number of national agricultural research institutions. These institutions are diverse in needs and situations, yet often slender in resources, compared with many of those focused on wheat and rice. The maize research establishment thus is more cumbersome and difficult to manage than that for wheat and rice.

Researchers will not have an easy task if developing countries are to succeed in meeting future demands for coarse grains at acceptable economic and environmental costs. In the demand scenario, consumption of coarse grains in developing countries grows at an average annual rate of 2.8 percent between the late 1980s and 2030, a three-fold increase. Some of the increase—perhaps 1 percent per year—could probably be met by bringing more land under coarse grains in Africa and Latin America, but most of the burden will have to be borne by rising yields.

Resource-management approaches

Research designed to increase yields by developing improved cultivars of plant was fundamental to the Green Revolution in wheat and rice production. There is a role also for research designed to achieve yield increases by modifying the physical environment in which plants grow. This latter research can be called farming systems research because it focuses on the farm operation as a whole.

Farming systems research covers an enormous range of farming and research activities including work that is increasingly designated as resource management research. The following examples illustrate types of resource management research that are especially appropriate for addressing coarse-grain yields under rainfed conditions. One concerns measures to increase conservation of soil moisture. The second describes research designed to increase yields on the acid, infertile soils that characterize much of the tropics.

Conserving soil moisture. Yudelman and Hillel (1988) assert that, in India and elsewhere in the less-developed world, engineering approaches to soil conservation have proved uneconomical for many farmers. As a consequence, researchers, and at least some farmers, have turned to alternative techniques to conserve soil moisture (Greenfield 1988, 146). Low-tillage or minimum-tillage techniques that leave much of the previous crops' residue on the soil surface as mulch and leave the soil in rougher aggregates than conventional tillage are being investigated by researchers and are being adopted by some farmers (Yudelman and Hillel 1988).

In India, results of pilot studies from watershed development projects indicate that plowing on the contour combined with vegetative bunding offers "scope for significantly improved water retention" (World Bank 1991b, 51). The use of vetiver grass in the bunding is argued to be especially promising. The techniques can both reduce the susceptibility of crops to drought and lengthen the growing season by holding soil moisture longer than do current practices. Coupled with shorter and more water-stress tolerant varieties of cereals, pulses, and other crops, these techniques have potential for "significant yield improvement." Moreover, the technology "is both easily replicable at the farm level and involves modest investment costs, the core of the new technology only involving the simple and low-cost practices of contour plowing and vegetative bunding, coupled with short-season varieties and some fertilizer" (World Bank 1991b, 52).

The evidence described here indicates that there are a number of practices that, if the economics is right, farmers can draw on to increase soil moisture and thus improve crop yields in areas where water is the principal limiting factor in crop growth. Much of the coarse grain production in developing countries is grown in such areas.

Although the technical feasibility of these water-conserving practices is well established, the fact that the practices have not been generally adopted by farmers is prima facie evidence that they are not generally economical. Research to overcome the economic barriers may have substantial payoff in higher coarse grain yields, especially in dry areas.

Continuous cropping on acid infertile soils. Traditional farming practices on the acid, infertile soils characteristic of much of the tropics result in low yields and severe loss of soil productivity after 2 or 3 years of cultivation. Research over the past 10 or 15 years suggests that, with adoption of new management practices, these soils can be much more productive on a sustainable basis. Buringh and Dudal (1987, 33) assert that:

Various experiments have clearly shown that many deep tropical soils can be permanently cultivated if an appropriate management is applied: The main problem of these soils is not irreversible hardening or laterite formation, as is often believed, but the aluminum toxicity that can be corrected by liming. A poor nutrient status is corrected by appropriate fertilization.

Olson (1987, 216) writes that several examples exist worldwide of situations where low productivity brush lands or sandhills have been converted to productive land by the introduction of fertilizers "and other appropriate agronomic practices." Olson does not say so, but some of the research to which he referred likely was that done by a team headed by Pedro Sanchez, a soil scientist who worked at North Carolina State University. Reporting on that research, Sanchez et al. (1982) stated that they had developed technologies that permit continuous production of annual crops on some of the acid infertile soils in the Amazon Basin. Studies in the region show that three grain crops can be produced annually with appropriate use of fertilizer. In the 8-1/2 years preceding the report, 21 crops were harvested in the same field with annual average production of 7.8 t/ha of grain. (In the U.S. Midwest, annual yields of maize currently average about 7.5 t/ha.) Sanchez et al. (1982) reported further that, under this regime of continuous cropping, soil properties improved. A number of farmers in the region were induced to try the production system on parts of their land. The first eight who participated obtained yields comparable to those obtained at the research station. Other more long-term field results elaborating the findings of this group are emerging (e.g., Sanchez, Palm, and Smyth 1990). The results provide new and substantive data underpinning an optimistic assessment as to effective and sustainable cropping of tropical soils.

Conclusion

The main theme of this paper is that, if the demand scenario is met at acceptable economic and environmental costs, it will be because the supply of knowledge farmers are able and willing to use increases enough to compensate for the insufficient supplies of land, water, and other natural resources. In this argument, knowledge is the key resource, and increasing the supply of it on economical terms to farmers is the key policy issue in achieving global sustainable agriculture. The argument rests on the notion that the supply elasticity of knowledge is greater than it is for land, water, genetic, and climate resources. Experience over the past 50 years provides empirical support for the notion. It would be a mistake, however, to assume from this experience that the supply of knowledge will be as readily increased in the future as it was in the past. The past growth of knowledge did not just happen. It was the result of a deliberate commitment of resources by public and private institutions and by farmers. It cannot be assumed that the future commitment of resources will be on the requisite scale or that the payoff to the resources will be as high as in the past. Whether the incentives of private institutions will be sufficiently strong and the political will and perception of need in public institutions sufficiently well-developed is problematical. And if the knowledge needed should involve major scientific breakthroughs, e.g., if expanding crop yields at the needed rate should require substantial increases in photosynthetic efficiency, then even a large increase in commitment of research resources may not be enough to do the job.

Finally, it is not obvious how investments to expand the supply of knowledge should be distributed among human capital, institutional innovation, and new technology. The discussion in the previous section concentrated on issues in development of new technology, but noted that knowledge embodied in people and institutions is also critically important. There is no reason to believe that the marginal returns to these three kinds of investment are the same. Indeed, a major question in devising a strategy to expand the supply of knowledge is how to allocate the necessary investments among people, institutions, and new technology. I provide no answers to the question here. My point instead is to emphasize that the issue of the scale of the needed investment in knowledge should be addressed simultaneously with the issue of the optimal distribution of the investment among the three forms in which the knowledge will be embodied. Clearly the problem of expanding the supply of knowledge on a scale and in a way required for a sustainable response to the demand scenario will present political, economic, and intellectual challenges of the first order.

Literature Cited

Alt, K., C. Osborn, and D. Colaccio. 1989. *Soil erosion: What effect in agricultural productivity?* Agriculture Information Bulletin No. 556. Washington, D.C.: U.S. Department of Agriculture, Economic Research Service.

Anderson, J. R., R. W. Herdt, and G. M. Scobie. 1988. *Science and food: The CGIAR and its partners*. Washington, D.C.: World Bank.

Anderson, J., and R. Herdt. 1989. The impact of new technology on food grain productivity in the next century. In *Agriculture and governments in an interdependent world*, ed. A. Maunder and A. Valdés, 683-694. Proceedings of the 20th International Conference of Agricultural Economists. Aldershot, England: Dartmouth.

Binswanger, H. 1989. The policy response of agriculture. In *Proceedings of the World Bank Annual Conference on Development Economics*, 231-58. Washington, D.C.: World Bank.

Buringh, P., and R. Dudal. 1987. Agricultural land use in time and space. In *Land transformation in agriculture*, ed. M. G. Wolman and F. G. A. Fournier, 9-43. Chichester, U.K.: Wiley.

CIMMYT. 1989. The wheat revolution revisited: Recent trends and future challenges. In *1987-88 CIMMYT world wheat facts and trends*. Mexico City.

Crosson, P. 1986. Soil erosion and policy issues. In *Agriculture and the environment* ed. T. Phipps, P. Crosson, and K. Price. Washington, D.C.: Resources for the Future.

Crosson, P., and J. R. Anderson. 1992. *Resources and global food prospects: Supply and demand for cereals to 2030*. Technical Paper 184. Washington, D.C.: World Bank.

Dregne, H. E. 1988. Desertification of drylands. In *Challenges in dryland agriculture: A global perspective*, ed. P. Unger, T. Sneed, W. Jordan, and R. Jensen, 610-12. Amarillo/Bushland: Texas Agricultural Experiment Station.

Eckholm, E. 1976. *Losing ground: Environmental stress and world food prospects*. New York: Norton.

El-Swaify, S., E. Dangler, and C. Armstrong. 1992. *Soil erosion by water in the tropics*. Honolulu: University of Hawaii.

Foreign Agricultural Service. 1990. *World grain situation and outlook*. FG 3-90 (March). Washington, D.C.: U.S. Department of Agriculture.

Greenfeld, J. C. 1988. Water conservation: Fundamental to rainfed agriculture in developing countries. In *Challenges in dryland agriculture: A global perspective*, ed. P. Unger, T. Sneed, W. Jordan, and R. Jensen, 145-6. Amarillo/Bushland: Texas Agricultural Experiment Station.

Hawkes, J. 1985. *Plant genetic resources: The impact of the international agricultural research centers*. CGIAR Study Paper 3. Washington, D.C.: World Bank.

International Economics Department. 1990. *Price prospects of major primary commodities*. Vol. 2, *Agricultural products, fertilizer, and tropical timber*. Report no. 814/90. Washington D.C.: World Bank.

Lal, R. 1984. Productivity assessment of tropical soils and the effects of erosion. In *Quantification of the effect of erosion on soil productivity in an international context*, ed. F. Rijsberman and M. Wolman. Delft, The Netherlands: Delft Hydraulics Laboratory.

Lal, R., and B. Okigbo. 1990. *Assessment of soil degradation in the southern states of Nigeria*. Environment Department Working Paper no. 39. Washington, D.C.: World Bank.

Larson, W., F. Pierce, and R. Dowdy. 1983. The threat of soil productivity to long-term crop production. *Science* 2319(4584): 458-65.

Mabbutt, J. 1984. A new global assessment of the status and trends of desertification. *Environmental Conservation* 11(2): 103-13.

McNeely, J. A., K. R. Miller, W. V. Reid, R.A. Mittermeir, and T. B. Werner. 1990. *Conserving the world's biological diversity*. Washington, D.C.: World Resources Institute.

Mitchell, D. 1991. Changing patterns of food and raw materials consumption in Asia/Pacific Region: Implications for primary commodities. Washington, D.C.: International Economics Department, World Bank. Photocopy.

Nelson, R. 1988. *Dryland management: The land degradation problem*. Environment Department Working Paper no. 8. Washington. D.C.: World Bank.

Oldeman, L. R., R. T. A. Hakkeling, and W. G. Sombroek. 1991. *World map of the status of human-induced soil degradation: An explanatory note*. 2nd ed. Wageningen: International Soil Reference and Information Center; Nairobi: United Nations Environment Programme.

Olson, R. A. 1987. The use of fertilizers and soil amendments. In *Land transformation in agriculture*, ed. M. G. Wolman and F. G. Fournier, 203-26. Chichester, U.K.: Wiley.

Pardey, P. G., J. Roseboom, and J. R. Anderson. 1991. Regional perspectives on national agricultural research. In *Agricultural research policy: International quantitative perspectives*, ed. P. G. Pardey, J. Roseboom, and J. R. Anderson, 197-264. Cambridge, U.K.: Cambridge University Press.

Parry, M. 1990. *Climate change and world agriculture*. London: Earthscan.

Peeters, J., and J. Williams, 1984. Towards better use of gene-banks with special reference to information. *Plant Genetic Resource News* (FAO) 60:22-32.

Pierce, F., R. Dowdy, W. Larson, and W. Graham. 1984. Soil productivity in the Corn Belt: An assessment of erosion's long-term effect. *Journal of Soil and Water Conservation* 39(2): 131-36.

Pingali, P. 1991. Technological prospects for reversing the declining trend in Asia's rice productivity. Paper presented at the World Bank Conference on Agricultural Technology, Current Policy Issues for the International Community and the World Bank, Airlie House, Virginia, October 21-23.

Plucknett, D. L., and N. J. H. Smith. 1988. *Plant quarantine and the international transfer of germplasm*. CGIAR Study Paper 25. Washington, D.C.: World Bank.

Plucknett, D., N. Smith, J. Williams, and N. Anishetty. 1987. *Gene banks and the world's food*. Princeton: Princeton University Press.

Ruttan, V. 1991. Challenges to agricultural research in the 21st century. In *Agricultural research policy: International quantitative perspectives*, ed P. Pardey, J. Roseboom, and J. Anderson, 399-411.

Sanchez, P. A., D. E. Bandy, J. H. Villachica, and J. J. Nicholaides. 1982. Amazon Basin soils: Management for continuous crop production. *Science* 216:821-7.

Sanchez, P. A., C. A. Palm, and T. Smyth. 1990. Approaches to mitigate tropical deforestation by sustainable soil management practices. In *Soils on a warmer earth: Developments in soil science*, ed. H. W. Scharpenseel, M. Schomarker, and A. Ayoub, 211-20. New York: Elsevier.

Stocking, M. 1984. *Erosion and soil productivity: A review*. Consultants Working Paper No. 1. Rome: AGLS, FAO.

Wilson, E. G. 1988. The current state of biological diversity. In *Biodiversity*, ed. E. O. Wilson and F. M. Peters, 3-18. Washington, D.C.: National Academy Press.

Wilson, E. O. 1989. Threats to biodiversity. *Scientific American*. 261(3): 108-16.

World Bank. 1991a. *India—irrigation sector review*. Vol. 1, *Main report*. Report No. 9518-IN. Washington, D.C.: World Bank.

World Bank. 1991b. *India—irrigation sector review*. Vol. 2, *Supplementary analysis*. Report No. 9518-IN. Washington, D.C.: World Bank.

World Bank. 1992. *China: Environmental strategy paper*. Vol. 1, *Main report*. Report No. 9669-CHA. Washington, D.C.: World Bank.

World Bank/UNDP. 1990. *Irrigation and drainage research: A proposal*. Washington, D.C.: Agriculture and Rural Development Department, World Bank.

Yudelman, M., and D. Hillel. 1988. Prospects for technological change in dryland agriculture: An economic perspective. In *Challenges in dryland agriculture: A global perspective*, ed. P. Unger, T. Sneed, W. Jordan, and R. Jensen, 591-7. Amarillo/Bushland: Texas Agricultural Experiment Station.

Managing the Living Soil for Human Well-being

RICHARD HARWOOD
Michigan State University

The objectives of sustainable soil management are to optimize soil quality—its utility for agricultural purposes—in both the intermediate term (5 to 10 years) and the long term (over 10 years). In the past, soil scientists emphasized soil conservation, that is, preventing soil loss. The scientific literature is replete with research on how crop cover, tillage systems, and contouring and terracing affect soil loss. But in a world of increasingly limited arable land per capita, reducing soil loss is not enough. It is vital to optimize soil productivity to feed growing human populations while preserving the natural resource base. Doing so will require not only conserving the soil, but also finding ways to maintain or improve the soil's ability to absorb and hold water for crop use, to recycle crop-available nutrients with minimal loss, and to provide an optimal environment for crop root growth. These are the major defining characteristics of agricultural soil quality. A soil management system for agriculture is sustainable only when it maintains or improves soil quality (Larson and Pierce 1994)

Any examination of global soil management issues must differentiate between the possible and the plausible. Assumptions must be made about social good, economic viability, resource availability, and feasible environmental balance. Using technologies existing today, the world's population probably could be fed and clothed without soil, but the cost in energy, materials, and capital would be so high that few societies could afford it. In fact soil suitable for crop production, along with water and sunlight, is a reasonably well-distributed production resource among broad masses of people, providing enormous social benefit. For the foreseeable future, soil culture will continue to be by far the most economical and appropriate medium for most economic applications of photosynthesis.

Appropriate management of soil is extremely specific to time and place. The geophysical environment (including type, slope, and uniformity of the soil and frequency and amount of rainfall), markets, input costs, labor-to-capital ratios, and a host of other factors determine what constitutes optimal soil management. But most agricultural soils are greatly underutilized for achieving high productivity. The soil quality that results from conventional farming practices often leads to low efficiency in the use of water and of naturally occurring and applied nutrients (National Research Council 1993b). Current research is showing that management of soil biota (flora and fauna) is the major determinant of the utility of soils for agricultural purposes. Better management of soil biota through manipulation of cropping systems improves soil quality by enhancing nutrient cycling, and it increases retention of excess nutrients and other substances that are potential pollutants. Such an approach enriches agricultural biodiversity and strengthens environmental protection. It is entirely consistent with conservation goals.

First, a few basic assumptions about the importance of soil productivity in meeting the looming crisis imposed on the earth's resources by continuing population growth:

- Increasing amounts of food and fiber will have to be produced on a declining agricultural land base over the next few decades. Achieving the needed increases in production per unit area will require that *greater amounts of nutrients flow from soil to plant*.
- In a more populous world, the interfaces between human-managed systems and the environment will be increasingly fragile. Society will demand that agricultural systems improve containment of production materials (nutrients, pesticides, crop and animal residues, and soil) within field boundaries and in the upper soil horizons so that they do not become pollutants of nonagricultural ecosystems.
- The requirement for pesticides will decrease (but not be eliminated) as biological alternatives become more available and as pest and predator and disease balances are better understood.
- As demand for agricultural products grows and as plant genetic potential is enlarged, nutrient availability and flow will increasingly limit crop production. The cycling of nutrients into agricultural fields from fossil sources (application of fertilizer) and from recycling efforts as well as the mobilization of nutrients within the field and recycling from plant and animal residues and wastes back to soil will be central to raising crop yields at reasonable cost.

Today's conventional agricultural systems have serious shortcomings. They recover, in crop growth, less than half the nutrients applied as fertilizers. They are extremely inefficient in mobilization of soil nutrients. They do not adequately protect the soil from erosion. They are inefficient in containing nutrients, particularly at high flow rates. Most important, they lead, more often than not, to a decrease in soil quality, as indicated by the soil's ability to infiltrate and hold water, to maintain particle structure for optimal root habitat, and to hold and recycle nutrients. Less-than-optimal soil quality raises production costs in the long term, lowers production potential, and accentuates production variability.

This is not a discussion, then, about low-productivity systems nor of zero inputs, nor is it limited in any way to organic agriculture.

Soil Biota: An Underutilized Resource

For the past several decades, modern crop production has focused on soil chemistry and physics. Farmers typically add water-soluble nutrient forms to the soil, and sometimes the applications are calibrated and timed to the precise needs of the crop in an effort to improve efficiency. Clay mineralogy—the ability of soil particles to affect nutrient absorption and release—and the water-soluble portion of soil nutrients have been all-important. The Green Revolution and every other high-yield program required adequate supplies of fertilizer nutrients for success. Fertilizer inputs are used to bring soil to an "adequacy" level of available nutrients, as determined by soil chemical analysis, then "maintenance" levels are achieved by adding an amount of nutrients equal to the amount removed in crop harvest. The approach sounds reasonable, but experimental results over several crop seasons show that plants seldom utilize even half of the nutrients applied to the first crop in the sequence. In high-yield crop production systems in many environments, the remainder of the nutrients contaminates ground water or surface water, or both (Contant, Duffy, and Holub 1993). In many U.S. agricultural areas, the content of soluble nutrients in the ground water exceeds the recommended limits for human consumption (National Research Council 1989, 98-119). In most developing countries, particularly in the tropics, the paramount agronomic needs are mobilizing the nutrients in highly weathered soils and achieving maximum benefits from often scarce and expensive fertilizer. Most tropical soils, as they evolved under high temperature and rainfall, have undergone chemical changes that have reduced both their nutrient content and their ability to hold added nutrients in plant-available form. Improving this quality is the key to attaining high crop productivity at reasonable cost. And having a large and active community of soil plants and animals (soil biota) is critical to good quality.

The activities of soil biota in agriculture have been known for over a century, with the first detailed account being given by Charles Darwin (Darwin 1882). Darwin, in fact, devoted the latter portion of his life and writing career to the study of earthworms.

In the early part of this century, a practitioner-based body of indigenous knowledge began to evolve in Europe and in the United States, based on managing soil organic matter and particularly the humus fraction (Harwood 1990). Soil biota were seen as being critical to that process. *The Living Soil and the Haughley Experiment* (Balfour 1976) characterizes much of the thinking of the day and outlines an approach toward management of soil fauna and flora. With the growing availability of inexpensive macronutrient fertilizers during the 1950s and 1960s, however, interest in managing soil organic matter waned. Although the maintenance of organic matter as one of the conditions of soil quality continued to be seen as important (Brady 1974), it was not considered to be particularly controllable (Miller 1990), and it was regarded as being related to total organic matter input to the soil (Koizumi, Usami, and Satoh 1992). Thus, the application of organic matter sources such as manure, compost, or organic wastes was thought to be essential for increasing total soil organic matter. It has commonly been believed (Lowdermilk 1953) that a maximum-production crop like corn, which is grown with heavy applications of inorganic fertilizers and achieves high rates of carbon fixation through photosynthesis, would maintain adequate levels of organic matter in the soil. Such crop plants accumulate large amounts of organic matter, which is eventually incorporated into the soil. It is now recognized, however, that long-term soil organic matter content is a product of all these factors, but the eventual equilibrium between organic matter input and loss is determined by tillage and, particularly, by the soil's clay content and type, temperature, and moisture content. Soil organic matter loss is greatest when tillage is frequent and temperatures and moisture are high. Under intensive agriculture in the highly weathered soils of the humid tropics, soil organic matter equilibrium levels are low, usually below 2 percent. But total organic matter content is not a good indicator of soil quality.

In the late 1970s, studies of North American organic agriculture began to reveal the importance of the active fraction of soil organic matter (Hendrix et al. 1990). The active fraction is the 3 to 6 percent of the organic fraction that recycles, largely through the action of soil biota, over a period of 3 to 10 years (Harwood 1990). Although organically managed fields usually have no higher levels of total carbon than well-managed conventional fields, unless they receive heavy applications of non-crop carbon forms like manure or compost, their apparent active fraction is 50 to 80 percent higher (Harris 1993). In organically managed fields, soil tilth and water infiltration, as well as water holding capacity, are markedly improved, with correspondingly less variability in yields during short periods of moisture stress. Carbon and nitrogen mineralization potential (the ability of the soil to break down organic matter into plant-available nutrient forms) are higher, indicating the soil's capacity to transform nutrients more rapidly to available forms—hence it has higher quality "fertility." Not surprisingly, the amount of soil microbial biomass is correspondingly much higher, as well (Table 1). In the 1988 corn crop, for example, microbial biomass nitrogen was 186 micrograms per gram of soil for the low-input rotation but only 86 micrograms for the conventional system.

The data in table 1 are illustrative of the basic principles of soil quality management. In 1987 the low-input rotation was in its seventh year, having had six different crop and cover-crop species in a 5-year rotation. The conven-

TABLE 1
Soil microbial biomass activity in low-input and conventional cash grain cropping systems, long-term experiment at Rodale, Emmaus, Pennsylvania.

Cropping system	N mineral- ization (µg/g)	Biomass[a] N (µg/g)	Specific mineral- ization activity[b]	Biomass N : soil N (%)	N mineral- ization (µg/g)	Biomass[a] N (µg/g)	Specific mineral- ization activity[b]	Biomass N : soil N (%)
	Corn 1987				Barley 1988			
Low input	13.6	182	0.075	5.1	6.2	186	0.034	5.4
Conventional	8.2	133	0.060	4.4	4.6	132	0.035	4.2
Significance	***	*	**	*	*	*	ns	*
CV (%)	7	8	5	4	11	9	9	10
	Corn 1988				Barley 1989			
Low input	11.6	186	0.063	5.2	4.6	188	0.025	5.7
Conventional	5.4	86	0.064	3.0	2.6	103	0.026	3.5
Significance	*	**	ns	**	ns	**	ns	**
CV (%)	25	3	33	7	26	9	26	9

Source: Harris (1993).

[a] Biomass is the weight of microbial organisms, expressed in micrograms per gram of soil. [b] N mineralization / biomass N

*, **, and *** significant at the 0.05, 0.01, and 0.001 levels, respectively; ns = not significant at p = 0.05.

tional rotation of corn-soybeans-barley had just these three species over the same period. The nitrogen mineralization potential for the 3 years shown in Table 1 averaged 73 percent higher for the low input (high crop diversity) rotation than for the conventional (low crop diversity) rotation. In the conventional system, annual applications of fertilizer were made while in the low-input rotation there was no nutrient input other than biological nitrogen from legumes. Microbial biomass nitrogen averaged 63 percent higher for the low-input (high diversity) systems. And most important, because it indicates ability to recycle nutrients from organic matter to plant-available forms quickly, the portion of soil nitrogen that was in the active (microbial) fraction of the organic matter increased from 3.5 to 5.7 percent, a 63 percent gain. In these same plots, soil tilth has increased markedly and water infiltration capacity has increased by 30 percent as a result of 7 years of crop differences. The increase in overall soil biotic activity, as indicated by the more easily quantifiable microbial biomass, thus greatly increases soil quality, as measured by favorable plant root habitat and ability to absorb and hold water and to recycle nutrients. The efficiency of recycling and retention of applied nutrients is crucial to both crop productivity and efficiency.

Many scientists have reacted skeptically to this information, saying that it is interesting but largely irrelevant, that organic agriculture is not practical on a global scale because there simply is not enough animal manure. The stage has been set, however, for a major breakthrough in agronomic thinking.

Soil ecologists now speculate that soil microbial activity is primarily governed by organic matter diversity rather than quantity, at least with average agricultural levels of organic matter input from crop plants growing in the field. At Michigan State University, we are finding higher levels of microbial biomass, and of microbial activity similar to those in the Rodale experiment, in *both organically and conventionally managed fields as well as in plots of native plant populations* that have not been tilled as compared with fields planted to continuous corn. That rotation effect means that the greater the variety of crop residues that form the organic matter, the larger the active fraction of the organic matter. Apparently the key is crop diversity, either within a field or over time (in rotation). There is difference of opinion as to the window of opportunity for temporal achievement of diversity. Some scientists think that a substrate (crop residue) effect lasts 2 to 3 years, as measured by corn following soybeans, while others believe that the effects from many crops persist for 4 to 6 years. At Michigan State University, we feel, as do scientists in the Tropical Soil Biology and Fertility Program headquartered in Kenya, that the effect reaches near-optimum levels when four to six crop species are grown within a 5-year period. These species may be either a series of commercial crops or commercial crops that are interplanted with cover crops as mixtures. In the latter case, the cover crop is overseeded into a standing crop before harvest, e.g., clover into wheat or annual ryegrass into corn.

A crucial question is how sensitive this effect is to different soils and climates and to disruption through tillage or application of chemicals or nutrients. Organic practitioners believe the effect is highly sensitive (Vogtman 1984), but our preliminary data show it to be quite persistent. It may be enhanced by applications of manure or compost, which could possibly substitute for in-field crop diversity, but they clearly are not necessary. If high levels of soil biota, and particularly microbial biomass, can be maintained, their contribution to soil quality and productivity can be enormous. If substrate (crop residue) diversity will by itself increase microbial biomass by 50 to 80 percent, irrespective of most chemical treatments, it can have great relevance to commercial agriculture. The factor of greatest economic and environmental importance, however, is long-term retention of applied nutrients and efficiency of recycling them. There are indications that such efficiencies are being achieved, but it is not known whether the effect stems from greater soil biotic activity or is a direct consequence of the crop diversity, or both.

While it seems to be increasingly clear that soil quality is highly dependent on crop diversity, it seems also clear that nutrient retention and recycling at high flow rates (high crop yield levels) are very much dependent on how plant diversity is arranged in both time and space. At Michigan State University, we feel that properly managed rotations, through their impact on soil biota, are the key to management of soil quality.

Design of Plant Communities for Better Soil Use

The widespread contamination of ground water and surface water by nutrients in the United States and Europe has given rise to considerable efforts to reduce nutrient loss and to improve efficiency of nutrient use. Fine-tuning (and reduction) of fertilizer inputs and better handling of animal manures and other wastes is increasingly emphasized using a wide range of approaches (Contant, Duffy, and Holub 1993). There is renewed interest in cover crops as a means for capturing residual nutrients (Hargrove 1991), irrespective of their potential for increasing the activity of soil biota (Honeycutt et al. 1993). In temperate regions, aggressively growing fall crops can recover much of the soluble residual applied fertilizer, at least in more humid areas where winter leaching and runoff accounts for most of the annual losses. In temperate regions, then, where farm sizes are large and mechanical power for tillage or herbicides for no-till farming systems are inexpensive, crop diversity is achieved primarily by growing different crops in sequence or rotation. The addition of cover crops to these rotations for soil conservation and nutrient retention will be increasingly important in areas that are sensitive to nutrient loading such as the upper Chesapeake Bay, Long Island, and much of Michigan. The forces that will drive farmers to adopt such crop diversity will be environmental pollution, first, and economic efficiencies, second. In tropical regions with small farms, similar environmental protection benefits and more efficient nutrient use can be obtained with more complex crop mixtures.

The effects of landscape-level diversity between fields are less pronounced, but under intensive study. At current U.S. commodity prices and production costs, such diversity is rarely economical even though rotations and their effects on improving soil quality are well known (Faeth 1993). Energy input per unit of output may be 50 to 65 percent less for rotation systems (Chou 1993), but corn and soybeans are more profitable and hence more widely grown despite their inefficiency, adverse environmental impact, and deleterious effect on soil quality. Recent analysis has indicated that policy changes such as reduction of price support programs for crops grown in continuous monoculture systems are needed to alter present market bias toward unsustainable production (Faeth 1993; Contant, Duffy, and Holub 1993; National Research Council 1993b).

In the more populous developing countries, crop diversity is the rule rather than a new phenomenon. Farms are small and farmers have limited resources for inputs. Crop are grown in diverse patterns, whose structure is governed by soil type, climate, power source for tillage, markets, input availability, and the nation's level of industrialization. Many such cropping patterns have been extensively reviewed recently in *Sustainable Agriculture and the Environment in the Humid Tropics* (National Research Council 1993a). In the humid tropics, the best lowland soil types, especially if planted to lowland rice or to estate crops where inputs are not limiting, may be managed sustainably despite limited crop diversity. Adequate soil quality can be maintained with crop specialization if environmental pollution levels are acceptable. Lowland bunded rice paddies are highly effective in containing production inputs. On the more common, less-fertile soils, a quite different approach, usually combining perennial and annual crops, is desirable when there is high population pressure. For example, agroforestry systems involve an extremely broad range of plant species that maintain a high crop biomass with long-term accumulation of nutrients in plant material. Such systems accrue biological "capital" over time to achieve often high and seasonally-dependable economic output (National Research Council 1993a). Such systems can be optimized under conditions of high labor availability and readily accessible markets. Land tenure and other social and economic conditions favorable to capital formation and accrual are as essential for biological capital as for other forms of business capital (Lynch 1992). Animal integration is a common and even essential component for the high productivity of such systems. In all these systems, crop rotations and mixtures must achieve high nutrient flow levels along with containment of potential pollutants. In developed economies as well as in highly structured small-farm systems in populous developing countries, the determinants for high soil biological activity and for efficiency are in place. While plant and crop structure are different, the basic principles are similar.

Scientific Focus on Ecological Process

A considerable change in scientific approach is under way as more and more researchers embrace a biological systems approach to agriculture. Scientists' efforts to combine ecological theory with traditional agronomy is leading to significant advances in both soil management and in systems approaches to weed control. These two knowledge systems have significant overlap with indigenous knowledge, not only of alternative agriculture but of the more ancient, traditional farming systems. The breakdown of scientific bias that favored an industrial rather than an ecological approach is now opening new avenues for greater efficiency of resource use. Some of the many integrative approaches now used include (Harwood 1992, Rhodes and Harwood 1992):

- a maturing farming systems perspective
- agroecology
- integrated pest management
- systems theory
- the effective planning of agricultural systems to strengthen rural communities and improve the quality of rural life

An increasing number of scientists are now approaching systems structure with basic hypotheses for function and organization. The hypothesis stated earlier concerning substrate diversity as a determinant of soil biological activity is a good example. With the coming age of agricultural biotechnology, tools for genetic manipulation are being rapidly multiplied. Scientists are becoming able to manage not only the genetics of crops and animals, but the genetic makeup of pest populations. The management of genes in pests for susceptibility to agricultural control measures will be essential to the eventual stabilization and equilibrium of pest populations at sub-economic levels. Likewise the management of the genetic makeup of soil flora and fauna will be critical to improving long-term soil quality and productivity. Gaining the ability to accomplish this through substrate management is an exciting prospect for agricultural scientists.

Creating an Enabling Environment

Having a set of appropriate technologies, by itself, is not sufficient for long-term soil maintenance. Short-term economic forces are nearly always resource-exploitive. It is essential that market forces that encourage longer-term efficiencies be combined with incentives for environmental protection. The social benefits from such favorable interactions between production and environment must increasingly be factored into agricultural production economics. In many parts of Europe, Canada, and some states and communities in the United States, a wide range of such benefits are transferred to agriculture, such as community purchase of development rights, tax incentives, and payments to maintain open space and to employ sustainable practices such as cover cropping (Contant, Duffy, and Holub 1993). Disincentives to crop diversity such as the government's requirement for farmers to maintain a historical base acreage in order to qualify for crop subsidy payments should be eliminated. This rule forces a farmer to continue growing a constant acreage of a crop in order to remain eligible for subsidy. Regulation may be a necessary dimension to soil protection, but it is hoped it will not be the central determining factor.

In developing countries, the maintenance of soil quality and productivity is often more a social, political, and economic issue than a technical issue. Control and access to land is a key problem, particularly in the most common fragile soil areas that are under heavy population pressure. Many analysts identify this issue as crucial to soil preservation (National Research Council 1993b).

Sustainable Agriculture Based on Soil Quality as an Evolutionary Process

While the biological, social, and economic structure of agricultural systems must evolve with changing social, market, resource, and capital availability conditions, the structure, occurrence, timing, and composition of the biological components can be arranged in many ways to optimize productivity and maintain the soil resource. That optimal structure evolves over time and space as national development proceeds and as changes occur in farm families, communities, and national and global conditions.

Soil quality is deteriorating and in many cases being irreparably degraded as a result of agricultural science's (and society's) narrowness of vision. Our regard for soil as an inert, expendable resource rather than as a complex habitat for living communities has greatly undervalued it as a resource in a world of increasing resource scarcity. Our approach to maintenance of soil productivity as a resource thus should be based on the very structure and function of farming systems rather than depending on application of costly, after-the-fact corrective measures. We will not reach a high production potential until the living nature of soil is better appreciated. Yet achieving such a potential will be essential to feed world population in the early decades of the next century.

Literature Cited

Balfour, E. B. 1976. *The living soil and the Haughley experiment.* New York: Universe Books.

Brady, N. C. 1974. *The nature and properties of soils.* 8th ed. New York: MacMillan.

Chou, T. H. 1993. Energy and economic analyses of comparative sustainability in low-input and conventional farming systems. Thesis, Michigan State University.

Contant, C. K., M. D. Duffy, and M. A. Holub. 1993. *Tradeoffs between water quality and profitability in Iowa agriculture.* Iowa City: University of Iowa.

Darwin, C. 1882. *The formation of vegetable mold through action of worms, with observation of their habits.* New York: D. Appleton and Co.

Faeth, P. 1993. Evaluating agricultural policy and the sustainability of production systems: An economic framework. *J. of Soil and Water Conservation* 48(2): 94-9.

Hargrove, W. L., ed. 1991. *Cover crops for clean water.* Ankeny, Iowa: Soil and Water Conservation Society.

Harris, G. H. 1993. Nitrogen cycling in animal-, legume-, and fertilizer-based cropping systems. Dissertation, Michigan State University.

Harwood, R. R. 1990. A history of sustainable agriculture. In *Sustainable agricultural systems*, ed. C. A. Edwards, R. Lal, R. Madden, R. Miller, and G. House, 3-19. Ankeny, Iowa: Soil and Water Conservation Society.

Harwood, R. R. 1992. Biological principles and interactions is sustaining long-term agricultural productivity. In *Proceedings of the regional workshop on sustainable agricultural development in Asia and the Pacific region*, 27-49. Manila: Asian Development Bank and Winrock International.

Hendrix, P. F., D. A. Crossley, Jr., J. M. Blair, and D. C. Coleman. 1990. Soil biota as components of sustainable agroecosystems. In *Sustainable agricultural systems*, ed. C. A. Edwards, R. Lal, R. Madden, R. Miller, and G. House, 637-65. Ankeny, Iowa: Soil and Water Conservation Society.

Honeycutt, C. W., L. J. Potaro, K. L. Avila, and, W. A. Halteman. 1993. Residue quality, loading rate and soil temperature relations with hairy vetch (*Vicia villosa* Roth) residue carbon, nitrogen and phosphorus mineralization. *Biological Agriculture and Horticulture* 9(3): 181-99.

Koizumi, H., Y. Usami, and M. Satoh. 1992. Energy flow, carbon dynamics and fertility in three double-cropping agro-ecosystems in Japan. In *Ecological processes in agro-ecosystems*, ed. M. Shiyomi, E. Yano, H. Koizumi, D. Andow, and N. Hokyo, 157-71. Tukuba, Ibaraki, Japan: National Institute of Agro-Environmental Sciences.

Larson, W. E., and F. J. Pierce. 1994. The dynamics of soil quality as a measure of sustainable management. In *Defining soil quality for a sustainable environment.* Special Publication 33. Madison, Wisconsin: Soil Science Society of America.

Lowdermilk, W. C. 1953. *Conquest of the land through 7000 years.* Bulletin 99. Soil Conservation Service. Washington, D.C.: U.S. Department of Agriculture.

Lynch, O. L. 1992. *Securing community-based tenurial rights in the tropical forests of Asia: An overview of current and prospective strategies*. Issues in Development. Washington, D.C.: World Resources Institute.

Miller, R. H. 1990. Soil microbial inputs for sustainable agricultural systems. In *Sustainable agricultural systems*, ed. C. A. Edwards, R. Lal, R. Madden, R. Miller, and G. House, 614-23. Ankeny, Iowa: Soil and Water Conservation Society.

National Research Council. 1989. *Alternative agriculture*. Washington, D.C.: National Academy Press.

National Research Council. 1993a. *Sustainable agriculture and the environment in the humid tropics*. Washington, D.C.: National Academy Press.

National Research Council. 1993b. *Soil and water quality: An agenda for agriculture*. Washington, D.C.: National Academy Press.

Rhodes, R. E., and R. R. Harwood. 1992. A framework for sustainable agricultural development: Synthesis of workshop discussion. In *Soil and water quality: An agenda for agriculture*, 107-20. Washington, D.C.: National Academy Press.

Vogtman, H. 1984. Organic farming practices in Europe. In *Organic farming: Current technology and its role in a sustainable agriculture*, ed. D. F. Bezdick, J. F. Power, D. R. Keeney, and M. J. Wright, 19-36. ASA Special Publication 46. Madison, Wisconsin: American Society of Agronomy.

Panel

PIERRE ANTOINE
Winrock International

The Green Revolution in Asia was successful partly because it operated in a controlled environment where water and the climatic variations were not really limiting factors. The challenge that we face right now is to feed hundreds of millions of people—mostly in Africa but also in some parts of Asia—who are living in environments that are much less controlled, and very much subject to climatic variations. They really operate in a dryland agricultural context rather than in an irrigated context or in agroecological zones where rainfall is reliable.

Most of those lands are located in Africa, north and south of the Sahara. And if you have looked at television lately, you know that there have been some major disasters in Somalia, partly due to political problems, but the same type of problems also exist in Ethiopia, across the Sahel, and in southern Africa in countries like Zimbabwe, Mozambique, and others where political stability is not really the main problem.

Recently I was at a workshop in France, where a woman from Burkina Faso reminded the audience that the middle class in Africa these days is composed of people who can afford three meals a day. Basically, what that tells us is that little cash is available for millions and millions of people in Africa. Where there is no cash, all the beautiful schemes that are based on scientific findings and a variety of economic options, and so on, do not apply because there are no options there. If you don't have cash, you won't buy inputs. If you don't have inputs, well, how do you take care of the resource base, the soil?

Wherever I go in Africa, I am told by the local farmers and by the local research scientists that there is a degradation of the soil base—not degradation in the sense of desertification because of climate change or obvious wind and water erosion, but a much more insidious type of degradation. It is the slow dissipation of chemical and physical properties important for crop production, which you don't measure on a daily basis. The mining of nutrients is observed by farmers all over, and there are a number of explanations. The fallows are decreasing in length because of the population pressure. Also, because of the population pressure, people have to go farther and farther to find new land to crop. And that in itself creates a major impediment in soil management. You can't talk to the people about using compost that they produce around their farmstead to try to preserve the chemical fertility of their soil if they have to carry it to fields 10 kilometers away.

Soils in the African continent, at least in the semi-arid zones, are much more fragile than the soils in the temperate regions. The average useful depth of the soil on the periphery of the Sahara is often about 20 centimeters. At 20 centimeters, you're hitting hard clay—a zone that is not easily permeable to roots. As a result, even if you have a chemical nutrients reservoir there, it is not available. Moreover waterlogging is common. The soil's upper layers in addition are generally sandy. If you lose 2 of the 20 centimeters every 2 or 3 years, you are soon left without soil.

Winrock has a major program of soil management at the community level, with the involvement of research institutions in many of the countries where we are present. And I am struck by the fact that in the past year, I have not gotten any answer to any specific questions I have made to try to resolve the main technical constraints of our soil management programs.

Likewise, whenever I travel overseas, people ask me questions because they think that I have answers. I'm a member of the scientific community in the United States and I'm supposed to have knowledge. But, the level at which we operate and do research and focus on success cases and so on has nothing to do with the African reality where you don't have the cash and where you don't have many management options.

We have almost no hard data on anything. At a recent workshop, I made the comment that Africa has plenty of technologies on the shelf, and that the main problem was not to develop new varieties but to get those technologies to the field. A colleague of mine wrote and asked, "Do you have any hard data?" No, we don't have any hard data, but every time we talk to the farmers, they tell us they have heard about some new variety. In some cases, you discover that they have stolen seed of some variety or bartered for it, but usually their main complaint is they don't have access to it.

Well, we don't have any hard data on the export of nutrients and the import of nutrients. We have to rely on the fact that the farmers say, "there is a decrease in the fertility, decrease of yield," and so on. But, we don't have enough to substantiate our claim, so what do we do? We don't have any idea on the real rate of adoption of technologies, which generally is an indicator of the quality and the viability of those technologies.

As I say, success stories generally apply to other places. I know the work of Pedro Sanchez that Pierre Crosson mentioned, but it involves the use of lime on acid soils. When the people don't have lime or money to buy lime, that success story goes away. When you don't have availability of credit, how do you really try to be innovative? And then when you are unable to create some incentives for the people to use some of the options that you want to promote, you are in trouble.

How do we adapt the extension message and how do we influence policies when we don't have the hard data? This is where that I believe an institution like Winrock has to hook up with some universities and get some graduate students interested in looking at the reality. It may be less publishable. It may be less basic research, but it may be extremely useful. I am glad to report that a number of volunteers or students now would like to participate in some of our project activities.

Let me finish with a remark on the future. I believe that it is essential to use a participatory approach at the farm level. I am convinced that if we don't use that approach, what we propose will always be" accepted," because people are polite, but it won't be sustainable. The pitfall in the participatory approach is that sometimes if the people don't know enough about sustainability and the long-term impact of mining of nutrients and so on, they are unable to look at the urgency of the problems that they must address.

I like to work with NGOs, but NGOs don't like to collect data. They don't like to do any quantitative work because they don't have time and it's not their mandate. So, they may be questionable partners, when you want to try to have a more scientific approach in some activity that you want to replicate. At the same time, the reason that the NGOs are popular in Africa, I believe, is not only because of what they offer, but because the public sector has been unsuccessful at delivering extension messages. And I think that the NGO, in a way, is the private sector of Africa and may be the most viable extension alternative, although it is far from perfect.

PAUL BROWN
Winrock International

I was especially interested in Dr. Harwood's remarks about the importance of animal waste for maintaining soil viability, probably because my work is dealing with animal waste. What we see in the United States is that animal wastes can serve as a resource to build the soils and to contribute to soil diversity, but they can also be a pollutant or a waste to be disposed of, depending on how you handle them. I think there are a number of factors that, as the animal industry has developed in the United States, have caused these problems to come about. And they may be things that we need to keep in mind as we do development activities in other countries.

One of the factors that affects the use of animal waste is concentration. In this country, we've gone from a system where the individual farmer operated a system that was labor intensive. He had a natural diversity. He used his animal wastes in the farming system. They were cycled right back into the soil. Now we're seeing more and more concentration of animal feeding operations. I would liken this to monoculture in forestry or crop production.

Within the poultry industry, we have some areas where serious problems are developing, as for example here in Arkansas. As the industry has concentrated in order to supply processing plants, our institutions and our systems of handling the food supply have changed. We've had concentration. There are similar problems with the feed lots in the West. Where you have concentration of animals, you have concentration of the waste. I think that we have to consider institutional, social, and policy issues when we're looking at this and try to develop a sustainable system.

Another factor that needs some intensive research is content of the animal wastes. They are a nutrient source, but when they are overapplied, they're a pollutant. So you have this balance that you have to work on. You have to ask when you apply animal waste to the land, what are you going to use as the factor that limits your application: the nitrogen, the phosphorus, or the potassium? For example, if you apply poultry waste to fields to supply the need for nitrogen, then you soon build up an overabundance of phosphorus.

This can become a problem. We need to focus research on how to stabilize these elements and how to be able to apply them in the proper ratios.

Also, I think we should look at economics. When you concentrate the animal waste and when you look at the economic value, which generally is measured in terms of the nutrient content, then it becomes economically difficult to transport the waste very far from the areas of concentration because of competition from chemical fertilizers.

We need to get people to understand that there are two levels of cost. There is the production and processing cost, but there is also an environmental cost. And, when you look at that, it becomes a policy or an institutional question of who will pay that environmental cost.

These are some issues that must be addressed with poultry waste that might help us move the waste to areas where it could be beneficially used.

One thing that we have run across in our project that I suspect is true in any areas that experience concentrations of animal wastes is the tendency to look at one industry at a time. In northwest Arkansas, for example, if you do an analysis on the soils that are available for utilization of just poultry litter, it would seem that there is adequate land available without transporting it long distances. But this also happens to be an area of high concentration of beef cattle, of swine, of dairy operations, not to mention the human population. So, you must take the total animal waste stream into account when you're considering a management system or policy decisions that affect the handling of these animal wastes.

CHARLES FULTZ
U.S. Soil Conservation Service

I, too, wonder where we're going in soil conservation. We've made progress in this country, as attested by the historic data on soil erosion rates. But we need to be careful about where we're going in regard to conserving soil by reducing erosion.

Recently, I went back and read a little from W. C. Lowdermilk's *Conquest of the Land through 7,000 Years*. Lowdermilk tells a story in that publication that goes something like this. He heard of an old man on a hill farm in the South, who sat on the front porch as a newcomer to the neighborhood passed by. The newcomer, to make talk, said, "Mister, how does the land lie here." The old man replied, "Well, I don't know about the land a-lying, it's these real estate people that do the lying." Lowdermilk goes on to make the point that, in a very real sense, the land does not lie, but it does mirror a record of what men have written on it through their management practices. I think we must keep that in mind even in this nation where we have seen a lot of progress being made.

Dr. Harwood's paper talks about the living soil. We do need to get beyond protecting the physical resource of soil and look at the soil from a living standpoint. And as we do that, we'll be able to improve soil quality. And as we improve soil quality, of course, we'll be able to help meet the need for increased production in the future.

I'm encouraged that more of our agriculture leaders in our universities and researchers and scientists are looking to the concept of the soil as being a living medium and not an inert resource to be mined. Until we all buy into that concept in our agricultural practices, we're not going to be able to improve the soil quality to the point that we need in order to be able to feed future generations.

I also was interested in Dr. Harwood's paper as it relates to the 1985 Food Security Act. We will have around 140 million acres of highly erodible land in conservation compliance plans that were written through conservation districts in the Soil Conservation Service. For the most part, the practices that are being applied to these highly erodible acres are ones that Dr. Harwood talked about: diversity of crops, the substrate, and crop rotation. It would be interesting to be able to measure the impact that these practices will have not only on reduction of erosion but on improving soil quality, and over the long run, on increased production or, at least, more consistent production in different climatic conditions.

So I'm encouraged by some of the things that it seems like we're doing right in this country. We need to focus on the fact that as we improve soil quality we're improving our water quality as well. Farmers are becoming more and more interested in processes. They understand these processes. They know more about the environment. And they understand that they will have to change their methods of farming to some extent.

Discussion

Raun: Pierre, would you have any comment on how countries in sub-Saharan Africa deal with the problem of mining of nutrients and, particularly, phosphate?

Antoine: When I refer to mining of nutrients, I am really looking at the whole soil picture: what and how much is lost on a yearly basis and what and how much comes in—especially any nutrient that is essential for plant growth. So, we're talking about possible losses through erosion, through leaching, through infiltration, or through export by crops. We don't have comprehensive hard data, except in some isolated places, on how much you export with natural processes or the crops, how much you bring back with the compost or other amendments, and so on.

I am not sure that the situation must be looked upon in a pessimistic way. There is a growing interest by local communities in using green manure, returning animal manure to the soil, keeping the straw on the field without burning it, using agroforestry, and so on. That would resolve the problem of nitrogen. This is where we have to collect some hard data.

Fortunately, there is quite a bit of rock phosphate available in a number of places. On a recent trip to Africa, I was surprised, in the middle of nowhere, to find a person mixing rock phosphate with his compost. It looks like a practice that is picking up in the Sahel, and it's also being done in other places. So I don't have that many concerns about phosphorus if the people start using the technologies.

My main concern is that the people don't have money, and they won't have money in the foreseeable future, to invest in buying some other inputs that are necessary. I'm not sure that those methods that we are trying to develop at the farm level, trying to keep and to return as much nutrients as we can, meet the levels of what is exported. That's where we need data. And we also must take into account mineralization rates and so on, which are very high in tropical or subtropical climates.

So, I believe technical solutions are possible, but there won't be any solution if we do not include at the same time the socioeconomic component—incentives for the people, access to credit, a rural infrastructure, and all the other material constraints that plague Africa.

Peterson: I would like to ask Pierre Crosson whether he was referring to investments in research and technology for the restoration of lands or to investments in technology to protect land currently in use. I think there's an interesting difference between those two.

I was concerned because much of his presentation was based upon U.S. data and, as he noted, that might not necessarily reflect the situation elsewhere. In many cases the United States is still benefiting, in terms of maintaining our soil resource base, from investments that were made in these types of activities by the CCC and the WPA during the Depression.

But I don't think we're talking just about soil erosion but about the productivity of the soil in general. Waterlogging and salinity aren't soil erosion, but they do affect the soil as a productive resource.

Also, I'm not quite sure that we have many choices. Often the degradation of the soil resource base is a local issue that's not just an economic concern but also we need to invest in soils to make other investments pay off. If you put a lot of money into developing an irrigation system and don't put money into dealing with salinity and waterlogging, you will lose that investment fast. Or if you're going to invest in new seed varieties or new productivity technology, you need the soil to make that work. It's clear that if we don't do something about micronutrients, the big investment in the Green Revolution is going to have a much shorter life span than we thought.

I think that, finally, as population does increase, we will be moving more and more into marginal lands. I don't think that existing technology in our use of marginal wetlands is sustainable, especially if we don't pay attention to protecting what we have. At a certain point, if we don't either protect what we have or bring back into production what we've lost, exploitation of new marginal lands will become so expensive that we're going to be forced into major recovery activities of land already degraded. The point is that it is more efficient to protect existing resources than it is either to develop new marginal lands or to restore lands already lost due to environmental degradation.

Crosson: I didn't mean to leave the impression that I was inferring from the U.S. experience that the land resource is not a problem elsewhere. In fact, soils in the tropical areas generally are believed to be more susceptible to erosion than those in temperate zones and, in addition, the productivity impact of erosion in tropical areas is greater than it is in temperate zones.

My point is that we know little, but we make a lot of assertions. We are also beginning to make policies to deal with the perceived land-degradation problem in the developing countries, where research resources are increasingly scarce, and we're not talking about getting needed increases, we're talking about losing resources. So being clear about the nature of the problem becomes critical in setting our research priorities. I made a distinction between reducing land degradation, or more generally natural resource degradation, as one way to go to increase output and investing in new knowledge about how to increase the productivity of those resources as the other.

With respect to investments in restoration of already degraded resources or halting current rates of degradation, my impression is that there's much more literature on current rates of degradation than there is on the current extent of already degraded resources and what would be involved in upgrading those.

Pat Peterson also emphasized that the problem of the soil resource is not limited to soil erosion. My reaction is that's the one that is most prominently discussed and is generally believed, as I read the literature and talk to people, to be more threatening than other kinds of land degradation. Considering these other kinds of land degradation doesn't change my hypothesis that investing in reducing them would have less payoff than investment in new knowledge about how to improve the productivity of those soils.

Pat's point about investments in land, I take to mean that reducing land degradation may in effect be complementary to other investments—more use of fertilizers, higher yielding varieties of crops, and so on. That strikes me as a good reason for thinking further than I had done about the problem of land degradation. It's clear that soils degraded by severe erosion are less responsive to fertilizer, for example, particularly if the waterholding capacity of the soil has been affected. It would follow that they likely would be less responsive to higher yielding varieties for the same reason. So that would certainly be something to look at.

Pat's final point was that with continued population growth, particularly in the developing world—and I would add to that the stimulus of increased per capita income raising demand for food—it's likely that farmers will have to move onto more marginal lands. If that is so, then we're in deeper trouble than I think we're in. Within the existing knowledge regime about how to produce in agriculture, we're not going to be able to bring much additional land into production without beginning to run into severe economic and environmental costs because that will involve land clearing and drainage.

Trying to respond to this demand scenario, in the absence of abundant new knowledge, will, I think, almost surely exact very high economic and environmental costs. For that reason, I would say that, to the extent that we have to move very far in that direction, we're losing the game. If we want a sustainable agricultural system—clearly we do—we're not going to achieve that by moving very far onto more marginal lands. If that's the only course that's available to us, then there's a real question whether we're going to make it.

Grove: You and many others feel that a loss of productivity is not much of an issue in soil erosion. What happens if you put off-site costs in the equation?

Crosson: That's quite a different issue. The consequences of soil erosion for the environment, or more specifically the off-farm damages, strike me as much more threatening to water quality or to the longer term viability of irrigation systems than the threat of erosion to soil productivity. I should have made that clear. But it gives me an opportunity to say something else. I did refer to the likelihood that managing institutional change to achieve a sustainable agricultural system may be more difficult than developing the new technologies that will be necessary to prevent natural resource degradation. The reason is that property rights to environmental resources are ill-defined, so no one has much incentive to manage them sustainably. Moreover, because of the absence of clear property rights, markets for environmental resources are weakly developed or nonexistent. Consequently when expanding agricultural production puts pressure on the environment and environmental costs begin to rise, the costs are not reflected in prices, so we have no clear idea of what they are, and no one has incentive to take actions to control them. In a fundamental sense, therefore, environmental problems arise because of institutional deficiencies. And, as the consequence, in dealing with those problems, we must deal with those institutional deficiencies. At this moment in the United States where we've done a lot more research and where we'd like to think we have a system that is responsive to emerging needs, nonetheless, we have severe problems of dealing with environmental issues, not just in agriculture but across the board.

So as we move toward expanding agricultural output, these emerging environmental costs are likely to pose a variety of problems in resource management but, in particular, institutional problems in dealing effectively with those environmental consequences.

Seckler: This session illustrates an important role for Winrock working in the resource environment field, especially with NGOs around the world—that is to help people think through the infinity of problems in this field to establish what are the important problems, what are the not-so-important problems, and, then, how to focus resources on the important problems.

I was particularly struck that Pierre Crosson points out that the world thinks that soil erosion is a major problem in Africa, while Pierre Antoine, I believe rightly, points out that it's not soil erosion but nutrient mining that's the major soil problem in Africa. I agree with Pierre Crosson that we don't know how much soil erosion there is in the world. But all the evidence that I've seen indicates that it isn't a very big problem, except in some local areas and in cases were downstream effects of sedimentation harm ports, hydroelectric facilities, and the like. But nutrient mining is a catastrophe for Africa. It also creates another problem by making people go into extensive farming methods. So, if there's more sorting through these issues, I think we'll perform our role much better.

3
Water Resources, Agriculture, and the Environment

Introduction

SANDRA BATIE
Virginia Polytechnic and State University

As a child I thought that water was something very special. I grew up in the Columbia River Basin of Washington State, where it rained only 7 inches a year. A rainstorm was met with as much excitement as snow storms in Miami. My brother and I would run out of our house to look at this miracle from the sky. The Columbia River Basin is highly developed, and many of the conflicts that we see today such as salmon-dams conflict have deep historical roots with which I am well acquainted. Indeed, the Columbia River Basin was destined to be the "Columbia Basin Authority" until the plans were shelved due to a public backlash to the Tennessee Valley Authority.

There's an old western saying that sets the tone for much discussion of water, "Whiskey is for drinking and water is for fighting." I grew up knowing that water is a important resource. It's important economically, ecologically, and politically.

Economically, water is important. We have heard how much of the high-valued agriculture of today is dependent on irrigation, access to water, and the infrastructure that accompanies water. The value of the land and the standard of living of people who are using the land and its products depend on access to water. It is also clear that people's health is dependent on access to clean drinking water—so many human diseases are waterborne.

Ecologically, water is important. The analogy that water is to the earth as blood is to the body is quite accurate. If you look at the water cycle and its complexity, the connections between surface water, ground water, and rain, we realize that water is the very sustenance of life itself. The in-stream flow values such as fisheries and recreation come from having access to clean water, and these in-stream flow values can be threatened by some agricultural uses. So there is a tension between the two uses. These tensions have both quantity and quality dimensions. Water's assimilative capacity of "residuals" forms an important element in our pursuit of higher standards of living.

Politically, water is a important resource. It is politically important because it is so important ecologically and economically. That is why water is for fighting. However despite water being a very valuable resource, there are many perverse institutional incentives for its use. In the American West, we use irrigated water at highly subsidized prices to grow subsidized crops. In the former Soviet Union, the Aral Sea is literally disappearing in order to irrigate cotton for a suspect international market. The market is suspect in the sense that it may not return anywhere near the kind of resources that the Aral Sea, if it had been protected, would have yielded in fisheries and other values. There's much to be done to improve institutions that manage water resources.

Finally let me add that as a board member of Winrock I am pleased to see how many of these dimensions of water are recognized and addressed in Winrock projects.

Designing Water Resource Strategies for the 21st Century

DAVID SECKLER
Winrock International

Water is a unique and even rather mysterious natural resource. It exists, like a Hindu God, in several distinct states of being, or phase states: a liquid, a solid (ice), and a vapor in the atmosphere. As temperatures change, water alternates among these three states—manifesting itself, as liquid, solid, or vapor.

The total global supply of water, in all three states, is essentially fixed. Unlike other natural resources, water is not destroyed in the process of use. As it

---

Acknowledgment. Jack Keller, a Winrock senior associate, reviewed this paper and contributed to many of the concepts. I am also grateful to Peter Rogers, also a Winrock senior associate, for reviewing the paper and making valuable suggestions for improvement.

is used, it is not used up. And water cannot be economically produced on a large scale. Thus the global supply of water today is nearly the same as it was when the Earth was formed 4.5 billion years ago.

The reason is that hydrogen and oxygen do not have a natural affinity for each other. They can be combined into the molecule H₂O only by the application of heat, but after this atomic ménage à trois is formed, the elements become highly attached to each other and can be separated only by the application of extreme pressure. Thus virtually all of the global water supply was formed in the steam atmosphere at the Earth's creation. As the globe gradually cooled, the steam condensed into liquid, and rainfall formed the oceans. As temperatures declined still more, some of the liquid water solidified into ice, thereby lowering the ocean levels. While the total amount of global water has been essentially the same since creation, the proportions in the liquid, solid, and vaporous forms have undergone large long-term variations as global temperatures have fluctuated. The ice caps have grown thousands of miles and receded, and the oceans have risen and fallen in accordance (Unesco 1978, 42-45; Kasting 1993).[1]

While the global supply of water is fixed, we naturally are interested in which state it occurs and where and when. From this practical point of view, perhaps the most important fact about water is that some areas of the planet are short of water while others have surpluses. These differences in water regimes largely determine the distribution of human populations on the globe and substantially affect the evolution of entire civilizations.

The demand for water is primarily determined by population growth and the growth of economic activities. The traditional demands for water have been in the agricultural, urban, and industrial sectors. Global use of water in these sectors has increased three times over the past 40 years (Postel 1992, 39), growing at a compound rate of 2.3 percent per year.

In addition to water demands from traditional users, however, a major new consumer has entered the global water scene. This is the environmental community, which is demanding enormous amounts of water for wetlands and wildlife refuges, scenic lakes, fisheries, recreational uses, estuaries, wild rivers, and the like. If all the global environmental demands for water were

---

[1] Some water is created in volcanoes and oceanic hot spots, while some atmospheric vapor escapes into outer space, but these are thought to be rather small amounts that roughly balance out. There does, however, appear to be a problem in balancing large losses of water due to chemical reduction in the photosynthetic processes of plants with the amounts of water created (Unesco 1978:45). Also, of course, the quality of liquid water varies enormously depending on salts and other pollutants.

added up, it is likely that they would exceed urban and industrial demands and approach the size of agricultural demands. This is particularly true of the "consumptive" use of water, which occurs when water is lost to a locality through evaporation into the atmosphere. Many of the environmental demands for water are for large, exposed surfaces of water (like lakes, rivers, and wetlands) that are subject to high evaporation rates.

While water demands are rapidly increasing, development of additional water supplies is slowing. The World Bank, for example, has reduced its investments in irrigation to less than half the 1978 level, in constant terms (Yudleman 1993). The real per-unit costs of adding irrigated area have probably doubled over this period. Together, those trends suggest that the rate of growth of new irrigated area may have fallen to 25 percent of the rate 15 years ago. Many countries, such as India, China, and Egypt, are rapidly approaching the physical and economic limits of water development, with little potential for increasing water supply to satisfy additional demands.

One does not have to be a prophet to foresee that growing water demand combined with limited water supply will create conflicts in the next century; many are boiling right now. Egypt, for example, has in the past threatened to go to war with riparian states in the Nile basin if necessary to protect water supplies. Egypt wants to build the Jonglei Canal in Sudan to augment the flow of the Nile into Egypt. But the canal would drain an immense part of the Sudd, a vast wetlands area that is a major sanctuary for migratory birds from Europe.

In South Asia, India and Pakistan nearly went to war over the distribution of Indus basin water supplies, and an agreement had to be more or less imposed by the World Bank. India now is using the Farrakka Barrage on the Ganges River near the border with Bangladesh to divert water in the dry season to the Calcutta region. During the dry season, the flow of the Ganges into Bangladesh has been decreased to half its level of 5 years ago and is expected to cease altogether within this decade. This is causing concerns about increased saltwater intrusion in southwestern Bangladesh and damage to the Sunderbunds, another internationally important wetland ecosystem.

In 1991, a Middle East water conference scheduled to be hosted by the president of Turkey was canceled through the direct intervention of the U.S. Secretary of State on grounds that it would exacerbate tensions in that sensitive region. The Aral Sea in Turkmenistan and the Lop Nor Lake in China have nearly vanished because of diversion of water for irrigation. The Paraná River in South America is being made navigable so that large areas in Paraguay and Brazil can be opened to agriculture. This area contains the largest wetlands in the world, and care must be taken to protect them as the region develops.

The list of potential water conflicts around the world today could be extended indefinitely, and these do not even consider the substantially increased water demands of the next century. Truly, as they say in the arid American West, "Whiskey is for drinking and water is for fighting over."

Intensified water conflicts in the next century are certain, but the design of appropriate strategies for alleviating these conflicts is far from obvious. The unique physical features of water make it an extremely complex and difficult resource to analyze and understand. Before we can get our strategies right in this field, we need to get our thinking right.

The Physical Setting of a Water Resource Strategy

The hydrologic cycle

Figure 1 illustrates the global hydrological cycle. The cycle is powered by the energy of the sun, which annually evaporates 456,000 km^3 of water from the sea and 62,000 km^3 from the land mass of the continents. Of the total 518,000 km^3 of water that is evaporated, 410,000 km^3 precipitates directly back on the sea. The remaining 108,000 km^3 that precipitates on the land is the annual freshwater supply of the continents. Subtracting the 62,000 km^3 of annual evaporation from the land leaves 46,000 km^3 in surface and subsurface runoff back to the sea (Ehrlich, Ehrlich, and Holdren 1977). This is equal to the amount initially evaporated from the sea and precipitated on the land, and it is the net renewable supply of fresh water for the globe's landmass.[2]

Although the 46,000 km^3 of net water supply for the land is only 9 percent of the total hydrological cycle, it is a large amount of water. Imagine a 1-kilometer-square column of water extending upwards for 46,000 kilometers. But about two-thirds of this water, or 32,000 km^3, quickly flows to the sea, mostly in floods, leaving only 14,000 km^3 of effective freshwater supply to the globe (Postel 1992, 28). Most of this water falls and flows naturally over the globe, according to topographic and other conditions. About 3,200 km^3 is controlled by human intervention of one kind or another and is purposely applied to various uses (World Resources Institute 1992, table 22.1). A small fraction of the water that flows almost immediately to the sea represents the ultimate potential supply of fresh water for additional development through regulation of

---

[2] The flux of this system varies enormously. The average turnover rate of atmospheric water is every 9 days. The turnover rate of water near the surface of soil depends on evaporation rates and varies from less than 2 weeks to over a year. The water in ice caps and deep aquifers may remain for thousands of years.

surface and subsurface flows. For example, if the fraction that would be potentially practical to develop were 10 percent of the unused flow, or 3,200 km^3, that would be about equal to the amount already developed. Considering that most of the 32,000 km^3 of unused water is from precipitation in areas that already are moist and in undeveloped polar areas, 10 percent may in fact be a reasonable estimate of the practical potential for additional water development on the globe.

It should be noted that such terms as water "development" and water "losses" usually refer to fresh water—water in a reasonably pure liquid state—and to its usefulness, or value, in terms of specific locations, times, and qualities for specific purposes. For example, precipitation in the frozen Arctic

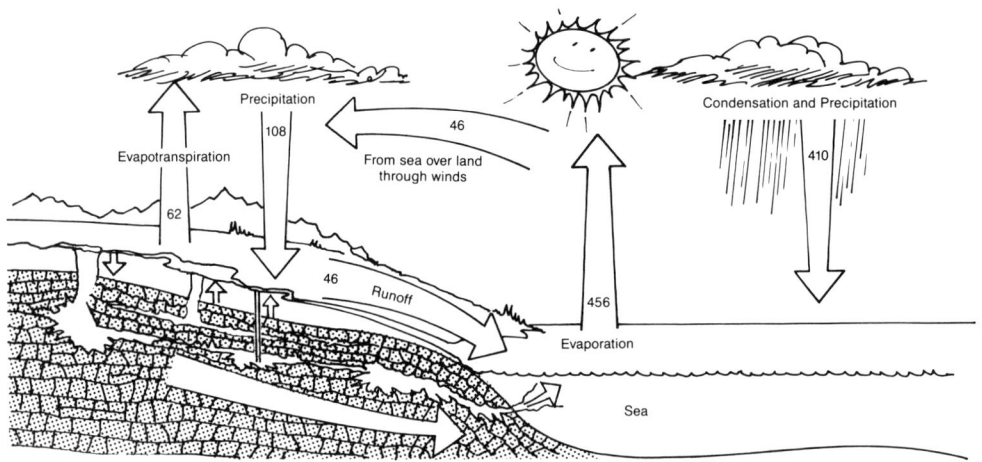

Fig. 1. **The global hydrologic cycle.** All numbers are in cubic kilometers of water per year. Adapted from From *Ecoscience: Population, resources, environment* by Ehrlich, Ehrlich, and Holdren, copyright (c) 1977 by W. H. Freeman, Co. and *Water resources atlas of Florida* by Fernald and Patton, copyright (c) 1984 by Florida State University Press. By permission of the publishers.

wastes may not be of direct economic value to humans, but it is important to the ecology of these areas and thus indirectly to humans. Similarly, other than for evaporation, the major loss of fresh water is caused by decreased quality from accumulation of pollutants or from excessive salinity when it mixes with sea water. But mixed fresh water and sea water is essential for maintaining estuarian habitats. And of course, saltwater can be recovered for other uses through desalinization technologies. In considering water, one must thus think in terms of a highly complex phenomenon that only can be defined in terms of many simultaneous dimensions. This is what makes water an intellectually intriguing subject, but also a very tricky one. Because it is easy to forget one or the other of these dimensions, it is easy to make important mistakes.

Water and heat

Another peculiarity of water is that its major biological use is not for water in any of its particular states, but rather in the transformation from its liquid and to its vaporous state: in evaporation. Use of water by plants probably accounts for more than 90 percent of the total water biologically used on the globe, and probably more than 98 percent of the water used by plants is used for controlling heat through the cooling effect of evaporation.[3] Most of the water consumed by animals is also used to control heat, although a large percentage is also used to dispose of waste products from the organism. Thus if temperature and precipitation rates were everywhere in balance on a continuous basis, all plants and animals would be in thermal equilibrium, with very little need for liquid water (except for habitats, waste disposal, and transportation) anywhere on the globe.

Evaporation and precipitation, however, are unequally distributed over the globe. Some regions of the world have higher evaporation rates than precipitation rates and thus are in water-deficit regimes, while others have the reverse relationship and are in water-surplus regimes. Some areas, like Egypt, would have large water deficits, but they receive supplemental water from upstream areas of their river basins that have surplus water. Some water-surplus areas, like most of Bangladesh, also receive additional water from their river basins, which creates major flooding problems.

Unfortunately, but logically, some of the largest areas of surplus water are in the polar regions, where evaporation requirements are nearly zero, precipitation is largely in the solid state of snow, and the biological productivity of plants and animals is low. Except for their environmental benefits, most of the large supplies of fresh water in these regions should be deducted from estimates of the actual and potential freshwater supplies available for productive uses over the globe.

Water is destiny

Figure 2 shows the influence of different water regimes on the distribution of the population of the globe. The shaded areas are those with positive water regimes, i.e., where average annual precipitation is equal to or greater than average annual evaporation, while the remaining areas have negative water regimes. The correspondence between population density and positive water

---

[3] Strictly speaking, evaporation from surfaces of the plant (transpiration) is "evapotranspiration," as opposed to evaporation from water and land surfaces. Evapotranspiration is used here only when plants are the sole object of the discussion.

regimes is quite remarkable. Perhaps over 80 percent of the world's population lives in regions with water-surplus regimes (or along major rivers flowing from such regimes). If we had a detailed view of the remaining 20 percent of the population of the globe, we would likely see that most of these people live in micro-watersheds that supply small streams and aquifers. The northern tip of Africa, which is fed by streams from the Atlas Mountains, is an example.

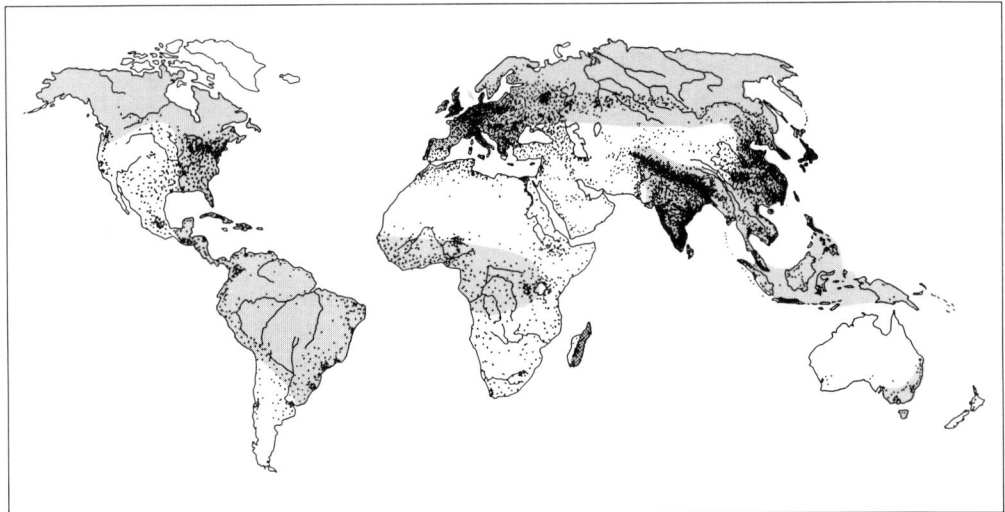

Fig. 2. **Global water regimes and distribution of population.** Shaded areas indicate positive water regimes—average annual precipitation is equal to or greater than average annual evaporation. Dots indicate areas of high population density. Adapted from Martin (1965), Glavnaia Geofizicheskaia Observatoriia (1972), Lydolph (1985).

Note: The dot population data was adapted from the *Encyclopaedia Britannica Atlas* (Martin 1965). The dots represent dense population areas: approximately 50,000 people per dot. The data is from 1940. This is the latest "dot" map we could find. Inspection of recent maps indicates that population density has increased in the same basic regions. Global water-regime data is produced from the maps of Glavnaia Geofizicheskaia Observatoriia (1972). Lydolph (1985) used the former as the source for his map. On Lydolph's map, the 0-mm line corresponds to the 400-mm line in the original source map. Lydolph does not explain this discrepancy. The shaded areas indicate where the difference between annual variation of precipitation and potential evapotranspiration is greater than or equal to Lydolph's 0-mm line. Mountainous regions are excluded from the presentation.

Most of the anomalies in the figure 2 are probably due to similar effects of runoff. The concentration of population along the major rivers, lakes, and coastal areas also illustrates the importance of water as a transport, trade, and fisheries resource.

Of course, figure 2 presents only a crude picture of water regimes. A region may have a positive water regime during certain seasons of the year, yet be in a deficit regime over the entire year. From an agricultural point of view, it is important in such cases for the positive season of the water regime to correspond with the crop-growing season. This is illustrated by the different water regimes for South India and South Korea in figure 3. As discussed below, wherever potential evapotranspiration exceeds effective precipitation, irrigation is required to achieve high productivity for most plants.

Figure 4 shows the variability of water regimes. Unfortunately, as indicated by comparing figure 4 with figure 2, there also is a high correlation between negative regimes and high annual and seasonal variability, and vice versa. Regions that are in or near a water-deficit regime, as is most of Africa, also are subject to higher variations in precipitation.

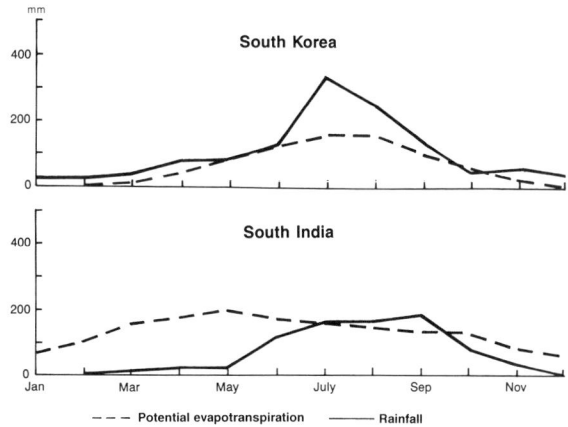

Fig. 3. **Seasonal moisture regimes in South India and South Korea**. Source: Wade (n.d.) after Thornwaite (1963).

The distribution of major saline areas of the world shown in figure 4 illustrates another important aspect of water, its service as a global cleansing agent. Once the H_2O molecule is formed, it is highly attractive to other elements, which tend to join it in a kind of extended family. For this reason, water is nearly a universal solvent. Also as water precipitates from the atmosphere it collects pollutants. This contributes to the pollution load of land and water surfaces, but water is highly effective in hastening the decomposition of many pollutants.

As precipitation falls onto, soaks through, and flows over the land, it picks up salts and leaches them to subsurface areas and the sea. These salts are of two kinds. First, *fossil salts*, as they may be called, are part of the mineral constitution of the planet. Agriculture would not have been possible unless water had flushed these salts from the great agricultural areas of the world. Second, *nutrient salts* are formed in the chemical reduction of plant and animal nutrients (whether organic or inorganic). These salts also must be leached from the soil to maintain plant growth. Again by comparing figure 4 with figure 2, it is seen that while not all dry areas are saline, most saline areas are dry. Major problems of salinity arise when these saline areas are irrigated, as is notable in the Indus basin of Pakistan, a former sea basin. A fascinating literature links the water-is-destiny theme to the effects of water regimes on the social, political, and economic evolution of civilizations. For example, in his great work on hydraulic civilizations, Wittfogel (1957) contends that the need to create and manage the massive irrigation works in the water-deficit regions of Asia led to centralized systems of "oriental despotism."

The water-surplus areas of Europe, on the other hand, especially those of

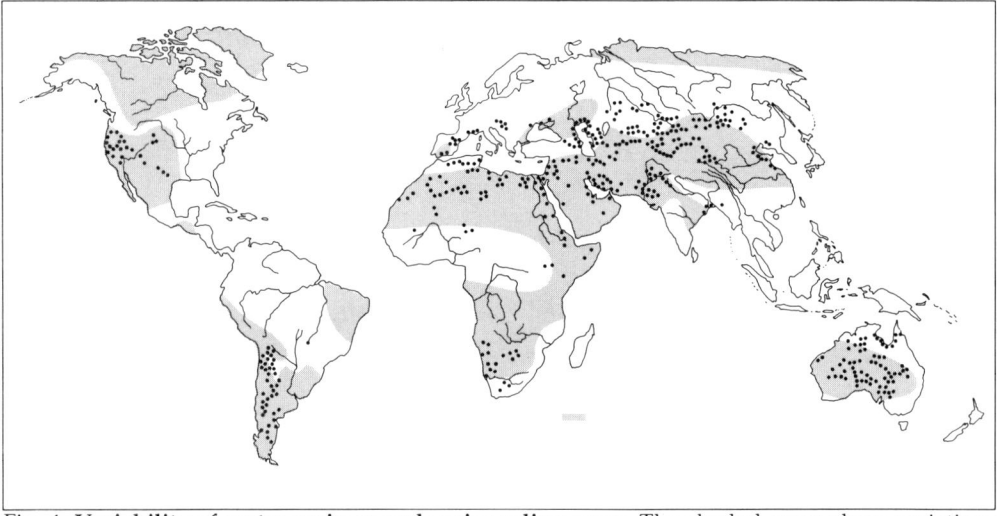

Fig. 4. **Variability of water regimes and major saline areas**. The shaded areas show variations of annual precipitation greater than 20 percent of normal; dots show important saline areas. Source: Fernald and Patton (1984, 5) and Snead (1980).

England, perhaps made it possible to achieve agricultural livelihoods in a highly decentralized, individualistic mode that contributed to the evolution of decentralized and more democratic forms of government. (Mountainous regions also tend to create this effect—hence, presumably, accounting for the rise of Athens, not to mention the insular ethnic chaos of the ancient Greek city states and the Balkans.) These effects are conditioned by the control of inland waterways, as in the feudal states of Europe, which are similar to the effects of controlling the overland and oceanic transportation routes by the great empires. Certainly, a review of figures 2 and 4 in light of the history of the regions provides circumstantial evidence for these historical water-regime theories.

Global climatic change

Last, but by no means least, water is the most important factor in regulating the globe's temperature. Among the many interactions between global water and temperature, it may be noted that the single largest source of uncertainty about global warming lies in lack of knowledge about the behavior of the oceans and atmospheric vapor (clouds) in response to and as modulators of temperature changes (Baskin 1993). Moreover, in terms of geologic time, before humans appeared on earth the globe was climatically warmer and more homogeneous than now. "[Man] came here in an ice age, and he has never known any other kind of climate.... The long periods of climatic geniality—never known by man—are called 'normal' times by the geologist" (USDA 1941, 7-8). A 2° C rise in global temperatures would restore us to a normal geologic climate. However, this could melt the ice in polar seas and Antarctica, which in the absence of off-setting effects would cause the ocean levels to rise substantially, inundating low-lying coastal regions like much of Bangladesh.[4]

While these characteristics of the global hydrologic cycle are intriguing, they may appear to be of little importance in the practical problem of designing water resource strategies for the next century. But the same concepts are fundamental to understanding the basic geographic unit for water resource analysis, the river basin.

River Basins

To use ecological terms, river basins are defined as areas with common and interconnected sources and sinks of liquid water. The common sources are annual precipitation and past precipitation stored in ice and aquifers. The common sinks are surface and subsurface flows of water to aquifers and seas (including oceans). Figure 1 illustrates the hydrologic cycle in relation to surface and subsurface flows in a river basin. As precipitation falls in a river basin, some part of it remains on the land surface, ultimately flowing to lakes and seas, while another part flows into aquifers, which are subsurface water basins. Some aquifers, like the Ogalala Basin of the western United States or those under the Sahara Desert, are *static* aquifers containing water that has been trapped underground, perhaps thousands of years ago. But most aquifers are *dynamic*—they slowly drain to rivers and lakes. (But it often is not known to what degree aquifers are static or dynamic, or the amount of water recharge

---

[4] However recent evidence, reported by Rensberger (1993), suggests that global warming mainly affects night and winter low temperatures. This would not melt the ice caps, but it could substantially increase agricultural productivity.

that occurs.) Water in dynamic aquifers may be trapped under an impervious layer of clay on a gradient, thus creating hydraulic pressure. When the gradient is right, simply poking a hole through the confining layer allows water to gush to the surface. Unfortunately, such artesian wells are as rare as they are remarkable.

The pressure in aquifers regulates underground flows at the interface of fresh water and sea water. If the pressure in the aquifer exceeds that of the sea, fresh water will flow into the sea; if the pressure of the sea is greater, saltwater intrusion will occur in the aquifer. One of the most important aspects of water

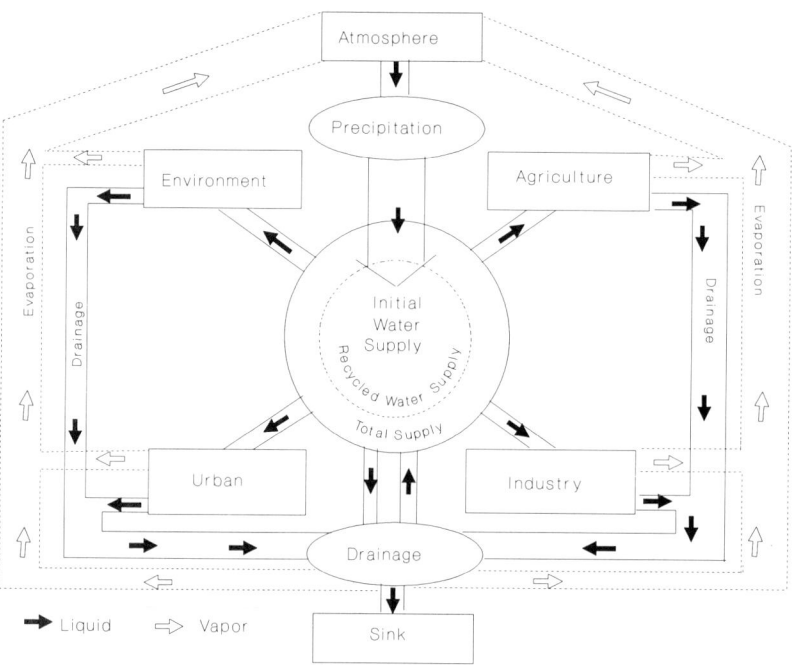

Fig. 5. **Hydrologic features of a river basin.**

resource management in coastal areas is maintaining the proper balance of these subsurface water pressures.

It often is thought that deep percolation losses of water are true losses in the sense that the water is never recovered. Fortunately, this generally is not true. Static aquifers eventually fill up and appear as lakes, or they can be mined by pumping. Some static aquifers, however, such as those in the Sahara, may be too deep to fill up (because of long-term climatic changes) or to allow economic recovery of water through pumping. Dynamic aquifers may drain to seas. But the loss of fresh water from this source is reduced because of back pressure from the seas. In general, except in deserts and some coastal areas, deep percolation losses are low.

The hydrologic cycle of river basins

Unlike the closed hydrological cycle of the globe, the hydrological cycle of a river basin is open. It receives water from outside the basin in the form of present and past precipitation, and it discharges water from the basin through evaporation to the atmosphere and flow to the sea. But within this basically open system, the liquid phase of the hydrological cycle of an entire river basin has water-recycling properties that are similar to those of the global cycle. These similarities increase as the river basin develops and becomes progressively more "closed."

The hydrologic features of a river basin are illustrated in figure 5. The annual supply of water to the basin provided by annual precipitation plus the water from past precipitation in aquifers and ice melt is called the initial water supply (IWS). Part of the IWS remains in rivers, eventually flowing to the ultimate sink, the sea. Another part is withdrawn from rivers and aquifers and applied to the major uses (agricultural, urban, industrial, and environmental) of water. As water is applied in these uses, part of it evaporates back into the atmosphere (shown as the open arrows), while another part drains away as liquid (solid arrows). Most of the atmospheric vapor from evaporation leaves the river basin, but some precipitates back into the same basin as a component of IWS.[5] Similarly, some of the drainage water flows to the sink while some flows back into the supply pool. Here it blends with the IWS and is recycled in the various uses. This is the "recycled water supply" (RWS). The sum of IWS and RWS is equal to the total water supply (TWS) of the basin.

As population intensity and economic activity increases in the river basin, a higher percentage of IWS is withdrawn by users, and a higher percentage of it is evaporated. But because only a fraction of the water applied in any given use is evaporated, more drainage water also is created. If this water is recycled, the TWS will grow. Thus, rather counter-intuitively, increased use and reuse of water in river basins increases the total supply of water for withdrawal and reuse within the basin. Specifically, if all the water users in a river basin evapo-

---

[5] The amount of evaporation that is recycled back to the watershed depends on climatic and morphological conditions, which vary enormously among watersheds. To my knowledge, no estimates have been made for this effect in any watershed. In discussing this question, Jack Keller suggested that the highly predictable afternoon showers of high mountain valleys may be due largely to this recycling effect. Afternoon showers in equatorial regions may represent the same effect.

rated the same percentage of the water applied in each use (E) and all of the drainage water was recycled, then the TWS of the system would be equal to IWS/E. If, for example, IWS = 1,000 units of water and E = 0.33, then TWS = 3,000. The ratio TWS/IWS, in this case equal to 3.00, is called the water multiplier (Seckler 1992).

But the practical limit to the water multiplier occurs before the point where all of the water is evaporated. As Keller (1992) shows, the water multiplier is stopped by the effects of another multiplier, the pollution multiplier. The initial water supply in a river basin contains some amount of salts and other pollutants. As the water is used and reused, it picks up more pollutants. These pollutants remain as the water is evaporated, so the concentration of pollutants in the water increases dramatically as water is reused. The concentration of pollutants is eventually so high that the water is not suitable for further reuse and is discharged to the sink.[6]

The water resource systems of river basins can be conceptually aligned along a continuum that varies with the degree of water recycling in the system. Near the origin of the recycling continuum are "open" water resource systems where water is withdrawn from the source and used only once. Part of the withdrawn water is evaporated to the atmosphere and the remainder is drained to the sink. The physical efficiency of water utilization in an open system is necessarily low. Even if all the water in the source is withdrawn for single uses, most of it will be drained to the sink. The irrigation efficiency of a typical surface irrigation system, for example, is less than 50 percent. This means that less than half of the water in the irrigation system is evaporated, the rest is lost to surface and subsurface drainage. Industrial and urban users use even less of the water withdrawn, perhaps 20 percent, so that about 80 percent is lost to the sink. In an open system, for example, if half of the water is withdrawn from the initial water supply of the system for irrigation and the other half goes to urban-industrial uses, only about 35 percent of the water would be consumed by evaporation and evapotranspiration, and 65 percent would be drained to the sea.

Naturally, as population and economic demands for water in an open system increase, more of the water flowing to the sink is captured and recycled by downstream users. As this process of reuse continues, the water resource system becomes progressively more "closed," moving toward the other end of the continuum. If water quality remained constant and there were no need to let

---

[6] Or fresh water can be recovered through desalinization techniques. The solid residue of salts and other pollutants then must be disposed of where they will not reenter the system, as in deserts or seas.

water enter the sink (e.g., to support estuarian habitats or prevent saltwater intrusion), a perfectly closed system would continue recycling water until all the water was evaporated. It would then attain 100 percent physical efficiency.

Studies of the nearly closed water system of the Nile (Abu-Zeid and Seckler 1992) and the more open system of the Imperial Irrigation District in southern California (Keller et al. 1992) provide examples of water resource systems that are at opposite ends of the recycling continuum.

The Nile River Basin

The Nile Valley of Egypt provides a nearly perfect case study of a water system in which there is extensive recycling. Because precipitation is negligible, there is only one input of water to the system—releases from the Aswan High Dam, which are carefully measured. As the water flows down through the Nile, it is diverted and used, and the drainage from each use either flows back to the Nile itself or to a shallow impervious aquifer from whence it is recycled. Finally, the surface and subsurface discharge of the Nile is reasonably well known. It is estimated that the water multiplier of the Nile below Aswan High Dam is about 1.85, i.e., all the water input to the system is used nearly twice on the average (Abu-Zeid and Seckler 1992). Because of this recycling, the system as a whole is operating at about 80 percent physical efficiency (i.e., about 80% of the water is evaporated as a result of beneficial use), even though the efficiency of every part of the system is much lower.

Further, as Keller (1992) shows in his study of the system, it would be possible to increase the overall efficiency of the system 10 percent, to nearly 90 percent, if salt pick up within the system could be reduced. The salt content of the initial water supply at the Aswan High Dam is about 250 ppm. With the current 80 percent system efficiency and no salt pick up, the concentration of salts in the drainage water should only increase about five-fold, to 1,250 ppm. But the average salinity of the drain flow to the sea is 2,800 ppm because additional salt is being picked up from saline deposits as the water cycles through the system. At this level of salinity, the water is not suitable for reuse by agriculture or other consumers and must be discharged to the sea. In addition, some amount of subsurface discharge into the Mediterranean is necessary to prevent saltwater intrusion into the Nile Delta.

As the Nile study shows, a closed water resource system may be operating at very high levels of efficiency even though all of the parts of the system are operating at low levels of efficiency. Of course, a high level of physical efficiency does not necessarily mean that the system is also at a high level of economic efficiency; that is a different question discussed below. It does mean that investments in improved physical efficiency (as in improving the efficiency of irrigation systems) may not yield a favorable return because there is little room for increased physical efficiency *at the level of the system as a whole*.

Of course, the Egyptian Nile is one of the most intensively used closed water resource systems in the world. Other systems, such as those of the Zaire and Amazon rivers, are almost wholly open. Between these extremes of the water-use continuum lie most of the remaining water systems of the world.

Water transfers in California

In open water systems, there is the potential for increasing both physical and economic efficiency of water use. In California, for example the Imperial Irrigation District (IID) is increasing the efficiency of water use in agriculture and selling the water saved to urban-industrial users in the area around Los Angles served by the Metropolitan Water District (Keller et al. 1992). The IID irrigates about 500,000 acres of land with about 3.1 million acre-feet of water from the Colorado River released from Hoover Dam in Nevada. The efficiency of irrigation in the IID is around 60 percent.

The IID is an open water resource system. Most of the water is used only once because the drainage water is highly saline, due to high the salt content of the soil. The discharge from the system, about 1,200,000 acre-feet per year, flows to a salt sink appropriately named the Salton Sea, which did not exist until drainage water from the IID created it in the early 1900s. The Salton Sea is highly saline and polluted with pesticides. It does, however, support some fish and waterfowl as well as recreational activities.

The IID has made agreements with the Metropolitan Water District to trade blocks of roughly 100,000 acre-feet of water in perpetuity for investments of about $100 million per block to improve the efficiency of the IID system. In other words, the water saved by improving water use efficiency will be transferred to the Metropolitan Water District. With high application efficiency, there will be less drainage water (and the remaining drainage water will have higher concentrations of pollutants).

Because this is currently an open system with 1,200,000 acre-feet of discharge flowing to a sink, this program will represent a real gain in physical efficiency, as well as a gain in economic efficiency through transfer of water to higher-valued uses (the Metropolitan Water District). The only major problem is that the sink, the Salton Sea, will now dry up, destroying its environmental uses. If these environmental uses are included, the system would be effectively closed, and net gains in efficiency would be achieved only in the economic dimension through the transfer to higher-valued municipal and industrial uses (if, indeed, they are higher-valued uses).

The consumptive use of water

As the Nile and Imperial Irrigation District examples indicate, there are two different ways of considering the physical efficiencies of water use. The first is in terms of the amounts of water withdrawn, diverted, or applied to each use cycle; the second is in terms of the amount of water evaporated in each use—or what is called the "consumptive use" of water. In closed water systems, where drainage water from a particular use is likely to be reused elsewhere, the consumptive use of water is the relevant criterion.

Switching from water-application rates to consumptive-use rates makes a considerable difference in how one thinks about the water efficiency of the various uses. A classic example is in hydropower production. While this sector accounts for fully 28 percent of the total water applied in the United States, its consumptive use, in terms of the additional evaporation it causes, is very low—probably less than 1 percent of total U.S. consumptive use. Of course, hydropower use of water may be costly in economic terms (in opportunity costs to other sectors and in timing of water discharges from dams versus the need to maintain high heads), but that is a different consideration.

The Colorado River basin of the western United States is one of the few water resource systems in the world that is explicitly recognized and managed as a closed system, with careful attention to return flows and water quality control. Accordingly, in this system, managerial and even legal decisions are based on the consumptive use of water and protection of the historic rights of downstream users to return flows of reasonable quality water. For example, owners of water rights in the Colorado River basin cannot sell or otherwise transfer their total water allotment, or water-application right, to another party. They only can transfer the consumptive use of water that is associated with that application right (Howe and Ahrens 1988).

In discussing water efficiency, Californians, in their inimitable way, distinguish between savings of what they call "wet" water and "dry" water. Improvements that reduce the consumptive use of water create wet water saving; those that only reduce application rates in a particular use are dry water savings (they simply reduce return flows to some other user). A later section shows how economic estimates in terms of consumptive use (i.e., wet water) substantially change the results of efficiency calculations on the basis of water-application rates, which may contain dry water.

Summary: Water multipliers as a paradigm shift

So far the discussion has been concerned mainly with the physical features of water resource strategies. The ultimate criteria, of course, are in terms of human values—the social, environmental, and economic dimensions of the strategies. But the physical features provide the basic set of parameters, the ultimate framework, in which the valuational criteria are applied. Thus in Egypt, there is a strategic choice between either improving the irrigation efficiency of specific irrigation applications, thereby reducing return flows from these applications, or letting the multiplier operate on high return flows and achieving high system-wide efficiency. This decision requires detailed analysis of relative costs and returns. In the lower Nile Delta, for example, where drainage is likely to escape to the sea, one might want to invest in high irrigation efficiency—but not in the upper Nile. Also, water recycling is likely to increase the salinity content of the return flow more than would high-efficiency irrigation (especially in more saline soil areas), at least in the short run. In economic terms, all these external benefits and costs of the system must be considered in developing appropriate water resource strategies.

In emphasizing the importance of water multipliers, the intent is not to close the discussion of options, but rather to open the discussion for considering a much wider variety of options. As these cases illustrate, the appropriate strategy for a particular water resource system and the potential for real gains in efficiency of water use in these systems largely depend on what kind of system one is dealing with—an open system, where the opportunities for physical gains are large, or a closed system, where they are small.

The distinction between open and closed systems and their associated water multipliers causes a kind of paradigm shift in water resources policy similar to that introduced in macroeconomics by John Maynard Keynes (1935).[7] As Keynes showed, what is true of every part of the whole is not necessarily true of the whole itself. Composition fallacies frequently arise in proceeding from the parts to the whole. Substantial mistakes can be made in thinking if analysis is not combined with—or even, one may say, under the control of—synthesis.

Water Conservation and Water Development Strategies

In March 1993, Winrock International and the Environmental Defense Fund sponsored a workshop with environmental and other groups on the future of water resources (Moore and Seckler 1993). One of the accomplishments

---

[7] In fact, the water multiplier concept was directly borrowed from the Keynesian income multiplier.

of the workshop was to clearly pose the differences between two basic models of water resource strategy, which may be described as the water conservation model and the water development model.

Proponents of the water conservation model argue that the present use of water resources is grossly inefficient, both in physical and economic terms. They contend that most of the world's future demands for water can be satisfied by water pricing and other forms of demand management—introducing more efficient technologies in irrigation, like sprinkler and drip irrigation; water conservation in urban-industrial sectors; and reallocating water from lower-valued (agricultural) to higher-valued (urban-industrial and environmental) uses.[8]

The conservation model can draw on a powerful example—the electric power industry of the United States. Two decades ago it was thought that massive investments would have to be made in power generation and transmission facilities to satisfy rapidly increasing future demand for electric power. The environmental movement, however, led a massive attack on this premise, arguing instead that conservation and improved efficiency of the existing system would generate sufficient surplus power to satisfy future demands. They were right. The United States has continued to grow economically over these 2 decades with little additional power generation. According to proponents of the conservation model, the situation with water is the same.

Proponents of the water development model (of whom I am one) challenge the basic assumption of the conservation model that water resources are now being used inefficiently. This, they contend, is merely an illusion created by concentrating on the obvious micro-inefficiencies of parts of the system while ignoring the macro-efficiencies created through recycling in the system as a whole. When power is used inefficiently, it truly is lost in the sense that it is dissipated in heat. But, when water is used inefficiently, it usually is not lost—it just flows to some other user. Thus proponents of the development model contend that most of the additional demand for water in the future must be met by development of additional water supplies. Water conservation alone will not suffice.

The question is not which of these typical models is intrinsically right or wrong. Rather it is a quantitative question of what mixture and what kind of water conservation and water development programs will contribute to meeting future needs. Some of the major factors involved in thinking about the answer for irrigation, the single greatest water user, are outlined here.

---

[8] Postel (1992) gives an excellent presentation of this thesis.

The case of irrigation

Irrigation accounts for around 80 percent of the global use of developed water supplies and even a higher percentage of the total consumptive use of water. It is commonly believed that because the agricultural use of water generally is inefficient, as in the California example, water can be diverted from agriculture to meet the demands of the other sectors without lowering agricultural production. But if, as in the Nile system, the physical efficiency of water use is very high at the level of the system as a whole, such transfers will reduce agricultural production.

Alternatively, it is thought that if a country encounters water shortages, it simply can take water out of agriculture, allocate it to higher-valued uses, and then import food to cover the shortage. But in addition to implications for food security, foreign exchange, rural employment, and other obvious aspects of this policy, there is danger from another composition fallacy. One or several countries can shift water from agriculture to higher-valued uses and import food. But all of the countries of the world cannot do this. As global water resource systems close and some countries reallocate water from food production and begin to import food instead, other countries will have to expand their agricultural production capacity, at least partly through development of additional water supplies, to generate food surpluses for export. It is likely that future comparative advantages in international agricultural trade will increasingly accrue to areas with water-surplus regimes, whether in irrigated or rainfed agriculture.

Some basic principles of irrigation

It may be helpful to review some basic principles of irrigation. The consumptive water requirements of plants is determined by two factors. The first is the potential evapotranspiration rate (ET_p) of the area in which the crops are grown. ET_p is determined solely by climatic factors (temperature, humidity, wind, etc.), which determine evaporation rates in an area (indeed, it often is measured by the evaporation rate of a pan of water in or near a field, calibrated for a reference crop such as vigorously growing grass). The second factor is the crop coefficient (k_c), which identifies the specific water requirements of different crops under the same ET_p conditions. The product of these two factors ($k_c \times ET_p$) defines the actual evapotranspiration rate (ET_a) of a particular crop, in a specific area, at a specific period of time, at its optimum, or maximum, yield potential.[9]

Once ET_a is defined, the water requirement of the crop (WR_c) is defined as $ET_a - EP$, where EP is the effective precipitation (the amount of precipitation that actually enters the root zone of the crop). In addition, WR_c is adjusted by the annual amount of water required to leach salts from the soil. The leaching requirement depends on the salinity of the soil and water. It may be zero in areas that receive substantial amounts of precipitation outside the crop season, as in most of Europe. But in dry areas, it can amount to as much as 20 percent of ET_a. Where ET_a is approximately equal to EP, the irrigation requirement is low or zero. This characteristic produced the great rainfed agricultural areas of much of Europe and the cornbelt in the United States.

Last, the irrigation requirement of the crop (IR_c), the amount of water that has to be delivered to the field, is determined by WR_c in relation to the efficiency of the irrigation system (EI):[10]

$$IR_c = WR_c / EI$$

In other words, IR_c is the amount of water that has to be withdrawn from the system (at the field gate) to irrigate crops. Under ideal conditions and with precision controls and facilities, all types of irrigations systems are capable of producing EI values in the neighborhood of 90 percent. However, EI typically varies enormously by the practical operation of the irrigation system employed. Typical surface irrigation system EI values range between 40 and 70 percent, while pressurized irrigation systems (sprinkler, drip, etc.) range between 60 and 90 percent. Unfortunately, in arid areas the need to apply more water than the crop requires, to satisfy leaching requirements, substantially reduces the efficiency gains of pressurized irrigation over surface irrigation precisely in those areas where efficiency is most needed.

We already have seen that in closed systems the efficiency of the irrigation system does not make much difference in terms of water supply at the level of the system as a whole because the inefficiencies are, as it were, recycled. Nor can anything be done about ET_p, which is exogenously determined by climatic factors. This leaves only the possibility of physical gains in system-wide irri-

---

[9] Economists naturally believe that improved efficiency is possible by reducing water-application rates to produce less than maximum yields, i.e., economically optimum yields. However, because precise water control in surface irrigation systems is lacking and the costs of mistakes are high, I do not believe there is much opportunity for gains in this direction (see Seckler 1992).

[10] Irrigation efficiency naturally varies at different levels of the irrigation system. At the "project level," for example, it includes losses in the conveyance system from the point of withdrawal at the head gates to the field gate.

gation efficiency from crop substitutions, by allocating water from crops with high k_c values to those with low k_c values.

Crop substitutions

Table 1 shows the k_c values of various crops in increasing order. Crops like safflower have the lowest k_c values, i.e., they require the least amount of water for maximum productivity under identical conditions of ET_p, while rice and alfalfa have the highest k_c values. The maximum amount of water saving among these crops, by substituting safflower for rice, would be about 30 percent. While this is a substantial potential savings, it ignores important factors.

TABLE 1
Crop coefficients.

Crop	Conditions		Crop	Conditions	
	Moist[a]	Dry[b]		Moist[a]	Dry[b]
Olive	0.40	0.60	Sugarbeet	0.80	0.90
Safflower	0.65	0.70	Citrus (weeds)	0.85	0.90
Grape	0.55	0.75	Cotton	0.80	0.90
Citrus (no weeds)	0.65	0.75	Green beans	0.85	0.90
Fresh pepper	0.70	0.80	Wheat	0.80	0.90
Groundnut	0.75	0.80	Dry onion	0.80	0.90
Green onion	0.65	0.80	Grain maize	0.75	0.90
Cabbage	0.70	0.80	Tobacco	0.85	0.95
Dry beans	0.70	0.80	Potato	0.75	0.95
Tropical banana	0.70	0.80	Fresh peas	0.80	0.95
Sunflower	0.75	0.85	Sweet maize	0.80	0.95
Watermelon	0.75	0.85	Sugarcane	0.85	1.05
Sorghum	0.75	0.85	Alfalfa	0.85	1.05
Tomato	0.75	0.90	Rice	1.05	1.20
Soybean	0.75	0.90			

Source: Hargraves and Samani (1986, 9).
[a] High humidity ($RH_{min} > 70\%$) and low wind ($\mu < 5$ m/sec). [b] Low humidity ($RH_{min} < 20\%$) and strong wind ($\mu > 5$ m/sec).

First, some crops are not agronomically substitutable because they grow under different conditions. Wheat requires cool weather, for example, while rice thrives in waterlogged areas that would kill most other crops, and it tolerates salinity better than most other crops. As a general rule, the more substitutable crops are, the closer are their ET_a values.

Second, crops with the highest k_c values also tend to be crops with the highest yields in terms of biomass, calories, or, sometimes, protein. For example, sugarcane yields the most calories per unit of land per unit of time, while alfalfa yields the most protein. Thus the physical biomass yield per unit of water consumptively used may be highest in the highest k_c crops. The reason is that, other factors being constant, both the yield of biomass and the water requirement of crops (WR_c) is determined by the surface area of the leaves.[11] The leaves intercept solar radiation, which largely determines biomass yields, and they transpire water. Thus, all things being equal, the greater the leaf area of the crop, the more yield and evapotranspiration. The major exception to this rule, as noted before, is in crops with different ET_p-growing regimes. For example, wheat and maize have the same k_c values, but wheat thrives in cool weather with low ET_p, while maize grows best with hot days (and cool nights) with high ET_p. Thus the reason why wheat uses less water than maize is not because of low k_c, but because it grows better in cool, low-ET_p seasons.

Third, the economic value of the crop yield is not included in these physical estimates. Traditional varieties of rice, for example, may yield only half as much as modern varieties, with more water use, but, because of taste preferences, traditional varieties may command a higher price in special market niches. Prices also are conditioned by the nutrient content of crops. In animal feeds, relative crop prices are almost wholly determined by their relative nutrient contents.

Fourth, all these considerations must be adjusted by the length of the growing season and crop rotation in relation to water needs, availability, and cost. Sugarcane, for example, has a growing season of 12 to 24 months, while rice, wheat, and maize require 3 to 4 months. Thus one might need to compare sugarcane with a rice-maize-wheat rotation over a several-year period to obtain valid results.

But all these complications aside, the basic principle is that in closed water resource systems, where drainage water is recycled, it is a mistake to estimate the economic returns to irrigation water by various crops on the basis of their irrigation requirements (IR_c), as it usually is estimated. Rather, estimates should be based on the consumptive use requirements of the crops.[12] Because the IR_c of rice, for example, can be over four times greater than its consumptive use, this can make a large difference in its imputed economic returns.

The sidebar (p. 93) shows estimates of net economic returns per unit of water consumptively used for a variety of crops in New Mexico. Crops such as

---

[11] This is true of total biomass production but not necessarily true of grain or fruit yield. L'vovich et al. (1990, 250) indicate an empirical confirmation of this generalization.

[12] Of course, this statement pertains to the "social," not the "private," economics of water. An individual farmer who has to pump water cares about the private costs of the irrigation requirement, not the fact that someone else benefits from his drainage water. It also is true, as in this example, that the costs of capturing and reusing drainage water are important from a social point of view—and these costs have to be considered in appraising the total system.

alfalfa and irrigated pasture, which have the highest consumptive use requirements, also tend to have the highest net economic return per unit of water consumptively used.[13] This would not necessarily be true of economic return to the irrigation requirements of these crops because their irrigation efficiency may be relatively low. But because of recycling drainage water in the basin, the irrigation requirement is not the appropriate criterion.

Water supply for irrigation

There is not much scope for improved efficiency of water use in closed systems of irrigated agriculture, but there is, of course, a considerable scope in open systems, assuming that they do not unduly conflict with environmental demands as they become more closed. Thus determining how much of the world's developed water supplies falls into closed and open systems clearly is an important area of future research. Estimates can be made by studying the discharge of the major river systems into their ultimate sinks. If we knew the quantity and quality of the discharge, especially by low-season flows, we could estimate the degree of closure of the system. This information, together with other information on aquifers, dam sites, terrain, etc., would then enable us to estimate the water development potential of the river basins in physical, economic, and environmental terms.

In my opinion, most of the important water resource systems of the world are nearly closed in the low-flow season, considering the need to maintain some discharge of pollutants from the system, to prevent intrusion of saltwater, and to sustain wetlands and estuarian ecosystems. If this is true, then additional water supplies must be provided by additional water development.

Although the use of water (in the sense of withdrawals) has tripled since 1950, it is doubtful that the development of new water supplies has increased this much. Rather, a good part of the additional use has been achieved by increased recycling of the initial water supply. Thus, as these systems close, the input of new water required per unit of water used may actually increase in the future, further exacerbating water shortages. However, the problem in estimating irrigation needs for the next century lies not so much on the supply side as on the demand side of the equation.

---

[13] The study described in the sidebar also suggests estimating the consumptive use of water in recreation and environmental uses. There is an indication that recreational values per unit of consumptive use may be lower than in most agricultural uses.

Sidebar: **Example of the Economics of Water Resources**

In a study of the upper Colorado River basin, Howe and Ahrens (1988) estimate the economic value of water in various uses to explore the possibilities for selling water among users. The Colorado basin is legally recognized as a closed system in the sense that only the consumptive use of water (as estimated by ET_a) can be sold legally. Irrigation water that is not consumptively used reenters the system, perhaps eventually to be captured and stored in Lake Mead behind Hoover Dam.

The table shows the economics of the crop systems practiced in the San Juan area of New Mexico in terms of the net returns to crops per unit of water consumptively used. It is clear that the crop system cannot be explained in terms of net returns per acre or per unit of water consumed. Other factors determine the system. But it is interesting to see (in the far right-hand column) that the crops with the highest returns to the consumptive use of water tend to be those with the highest water requirements, notably pasture and maize grain.

If one were to consider transfers on the basis of water withdrawn for irrigating these crops, with typically low efficiencies of irrigation, the value in agriculture would be only one-half, or less, of the amounts shown. This table shows how evaluations should be done in closed systems. In open systems, such as the Imperial Valley, where water is truly wasted, evaluations based on application rates are more appropriate.

Also, it may be noted that Howe and Ahrens (1988, 196) find a value of water for stream fishing, based on willingness to pay during low flow periods, of $11.47 per acre-foot. While they do not state what the consumptive use of the water is, keeping an acre of stream surface running for 3 months during the summer would cause consumptive losses about equal to ET_p, slightly above agricultural crop requirements. The recreational value per acre-foot of consumptive use thus would be about one-third as much as the upper end of the crop values in the table.

Crop water economics in the San Juan sub-basin of the Upper Colorado River basin, near San Juan, New Mexico, 1982.

Crop	Irrigated area (acres)	Annual yield per acre	Price ($)	Revenue ($/acre)	Production cost ($/acre)	Net return ($/acre)	Consumptive use Total (acre-ft/yr)	Consumptive use Net return ($/acre-ft)
Alfalfa	58,828	3.08 t	62.08/t	191.21	143.74	47.47	1.57	30.23
Barley	601	50.00 bu	2.46/bu	123.00	119.03	3.97	1.30	3.05
Wheat	1,716	50.00 bu	3.40/bu	170.00	138.72	31.28	1.67	18.73
Oats	1,738	50.00 bu	1.70/bu	85.00	69.35	15.65	1.60	9.78
Maize grain	256	87.64 bu	2.64/bu	231.37	158.30	73.07	2.08	35.13
silage	2,309	11.80 t	19.75/t	233.05	219.73	13.32	1.80	7.40
Potatoes	10	90.25 cwt	3.20/cwt	288.80	283.36	5.44	1.83	2.97
Pasture	65,200	6.80[a]	12.27[b]	83.44	24.58	58.86	2.00	29.43

Source: Adapted from Howe and Ahrens (1988).

[a] Animal unit-month. [b] Per animal unit-month.

Future demand for irrigation

The demand for irrigation is ultimately derived from the demand for food. The task of estimating world food demands in the next century has been enormously complicated by major structural changes in the global food economy. In an important study, Crosson and Anderson (1992) project that world demand for cereals will nearly double by 2030. About 90 percent of the increase will be in developing countries, mainly in Asia. The composition of increased cereal demand is expected to shift dramatically from the traditional foodgrains, rice and wheat, to feedgrains such as maize, barley, and sorghum. The reasons for this are the rapid growth in per capita income in Asia, the low or even negative income elasticity of demand for foodgrains, and the high income elasticity of demand for meat and other livestock products.[14]

Logically, the increased global demand for cereals can be met through a combination of four distinct paths—growth of cereal *yields* on irrigated or rainfed lands and growth of irrigated or rainfed cereal area. Unfortunately, the world agricultural production data, compiled by FAO from national data, does not separate crop area, yield, and production by irrigated and rainfed areas—the national agencies just lump them together. Thus, it is impossible to know on the basis of the statistics how much rainfed and irrigated agricultural lands have produced in the past or what might be expected in the future. This grievous problem in one of the world's most important data sets should be corrected. But in the meantime, all one can do is make guesses about the relative contributions of irrigated and rainfed agriculture to food production in the past and what it might be in the future.

One stylized fact that can be used in approaching this problem is that about 80 percent of the total increase in cereal production in Asia since the 1960s, over the period of the Green Revolution, occurred on irrigated land. This was due both to rapid growth of yield on existing irrigated land and to rapid expansion of the area of irrigated land. Now, it is generally believed, the rate of growth of yield from irrigated land in Asia has slowed substantially, falling to perhaps less than half its previous rate. At the same time, the growth of irrigated area also has decreased to less than half its previous rate. In addition, it is known that the total rainfed area of Asia and the world have been decreasing since the late 1970s. The situation with respect to cereal yields on

---

[14] It appears that these projections are close to the maximum cereal demand, and the actual demand may be significantly lower (Seckler 1993). However, the demand for irrigation for vegetables, fruits, sugarcane, and other food and nonfood crops may rise correspondingly.

rainfed areas is virtually unknown, but if yields are increasing at all in rainfed areas, it is at only modest rates.

Another loose end in assessing the food supply outlook is the vast areas of potential rainfed cropland in the interior of sub-Saharan Africa and Latin America, as discussed in Crosson and Anderson (1992). The costs of developing infrastructure for much of this land and the environmental damage this development may cause make the likelihood of realizing the potential doubtful.

Table 2 shows the existing and potential irrigated area of major regions of the world. My judgment is that these and other factors, combined with the natural desire of most countries not to become too dependent on food imports and attendant foreign exchange costs, will mean that by 2030, Asia, the Near

TABLE 2
Currently irrigated land and other land with irrigation potential.

Region	Irrigated (000 ha)	Potentially irrigable (000 ha)	Potential increase (%)
Developing countries			
Africa	11,025	18,175	165
North	7,560	1,640	22
Sub-Saharan	3,465	16,535	477
Western Hemisphere	16,235	22,865	141
Mexico and Central America	7,035	2,865	41
South America	9,200	20,000	217
Asia	158,380	69,420	44
Near East	18,315	5,185	28
Far East	140,065	64,235	46
Total	186,000	110,500	59
Developed countries			
Total	68,000	27,000	40

Source: Adapted from Crosson and Anderson (1992).

East, and Africa will need to develop their irrigation potential close to the limits shown in table 2. The developed countries (and much of Latin America) will not require further irrigation development. Considering recycling and water conservation potential, perhaps the global supply of water for irrigation will have to be expanded by only one-third, by roughly 60 million hectares. But this amount would appear to be the minimum required.

Further, within irrigation development, I believe that most of the increased production will need to be from large storage and conveyance systems. While small-scale water conservation techniques such as terraces, land leveling, and water-harvesting reservoirs can be of enormous local benefit in rainfed areas (Seckler and Joshi 1982), they will not substantially contribute to the task of doubling world food supplies. The reason is that water conservation techniques depend on local precipitation—and the amount and variation of local precipitation is the basic problem in areas of water-deficit regimes. Soil-moisture conservation typically can provide only up to 2 weeks of water storage in the soil. Small water-harvesting dams can provide only a few weeks of storage. These devices can modulate droughts, but they cannot prevent them. Worldwide experience shows that a reliable water supply is essential for inducing investments in labor, fertilizer, and other inputs necessary for a productive and sustainable agriculture. Large water storage and conveyance systems move large amounts of water in time and space, and that is what is required for substantial and reliable agricultural production in areas of water-deficit regimes.

Clearly, one of the most important areas of research for the future is in understanding the economic and environmental trade-offs between intensive and extensive production on existing and new irrigated and rainfed lands. These categories exhaust the range of possibilities for additional food production. Therefore, unless one is prepared to let the food situation worsen in the future, one cannot be against any of these alternatives without implicitly being for one of the others, or, as some people seem to be, against all these alternatives. You cannot, in other words, be against irrigation and fertilizer and also against expansion of agriculture into forests, grasslands, and wildlife habitats, and yet want to provide food for the rapidly increasing population of the world. Whatever the combination of choices may be, water regimes (whether through precipitation or irrigation) will play an ultimately decisive role.

Toward a Water Resource Strategy for the Next Century

A provisional outlook concerning water demands by the major sectors over the next 40 years may be advanced along the following lines. First, the industrial and urban sectors have the most rapid population and economic growth. Fortunately, these also are the sectors in which opportunities for increased efficiency of water use holds the greatest promise. More efficient use of water and more intensive water recycling can substantially reduce the need for additional water in these sectors.[15]

The consumptive use of water in these sectors can be reduced to perhaps 20 percent or less of the application rates. This would result in a water multi-

---

[15] In the terms used here, it may be said that often the water resource systems in these sectors are "closed" because the discharge water is polluted, but they may be "opened" by reducing pollution, as in the California example.

plier for this sector (including recycling effects to other sectors) of 3 to 4. The important thing in these sectors is to ensure that the discharge water is sufficiently clean to be recycled and that it is recycled, not simply dumped into a sink. Unfortunately, most of the population and economic activity in these sectors is located in low-lying coastal areas where it is easy to dump drainage water into the sea and difficult to pump it to where it can be recycled. Urban-industrial areas should be at the heads, not at the tails, of river basins from this point of view.

Figure 6 shows the costs of various water recovery methods. It is notable that the costs of treating brackish water supplies, potentially a very large source of water for urban-industrial users, have decreased to the level of conventional sewage treatment technologies.

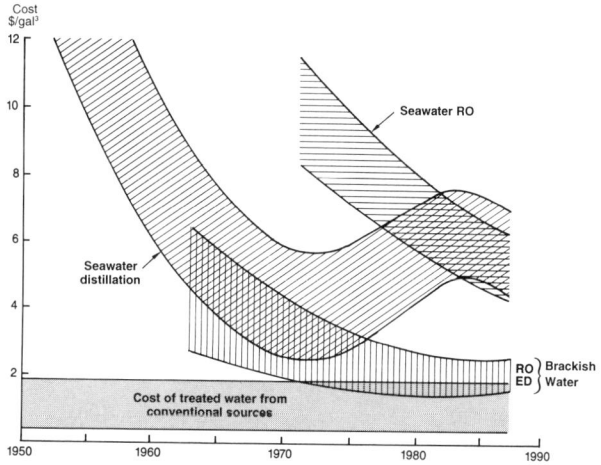

Fig. 6. **Trends in costs of desalinization, 1950-1990, in 1965 dollars** (RO = reverse osmosis; ED = electrodialysis). Source: Frederick (1993, fig. 1).

Even so, the urban and industrial sectors will require a considerable amount of additional water in the next century. Assume, for example, that population and economic growth in these sectors causes the demand for water to increase at a compound rate of 3 percent per year for the next 40 years; demand then will be 3.26 times its present level. But perhaps half of this increase can be met by water conservation, resulting in an increase of 163 percent in consumptive use in these sectors. Assuming they now use 20 percent of the world's total consumptive use, this would require a 32 percent increase in the world's total supply of developed water (or allocations from the environmental and agricultural sectors) to meet the demand for these sectors.

However, an extremely important aspect of the urban-industrial water sector is the massive economic costs of providing water and sewage treatment facilities for the rapidly growing urban populations in developing countries.[16] The sewage problem has largely been ignored so far. But as the concentration of pollutants increases to toxic levels, the problem will have to be confronted. Most of the growth of population in the next century will be in urban areas. It costs about US$500 per person to provide reasonably adequate domestic water and sewage facilities in urban areas of developing countries. Certainly, this will be one of the largest components of developing country investments in the next century. The economic aspects of urban water supply are far more important than the physical quantities would suggest.

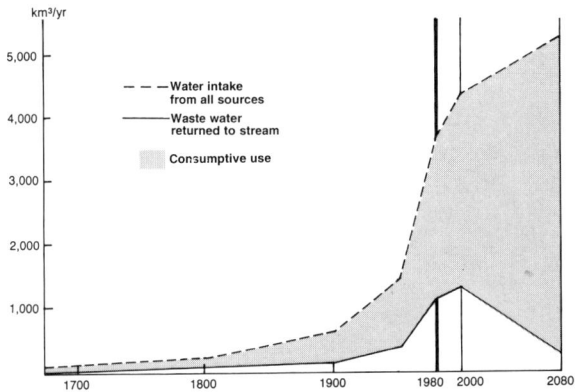

Fig. 7. **Chronology of water consumption for 400 years.** Projections from 2000 to 2080 assume drastic measures are adopted to reduce wastewater discharge (shaded area shows consumptive use). Adapted from L'vovich and White (1990, 248) (Copyright (c) Cambridge University Press 1990. Reprinted with permission of the publisher.)

Second, the consumptive use of water in the environmental sector is high because of high evaporation losses from large water surfaces and losses of water in streams to sinks. Perhaps the greatest single consumptive use in this sector is in wetlands, where the ratio of water surface to depth is high, and thus most of the water is evaporated. This is why a wetland area near Timbuktu in water-starved Mali evaporates nearly half of the entire flow of the Niger River. These consumptive losses should be included as part of all environmental and resource planning efforts.

---

[16] I am grateful to Peter Rogers for making me aware of the economic magnitude of this problem.

The large and increasingly intense demand for environmental uses of water, however, is mainly for preservation of water in its current use—not, usually, demand for additional water supplies. Thus these demands will be expressed mainly through protecting water from allocation to other uses. In this sense, future environmental demands represent more of a water supply constraint to other sectors than demand for an additional supply of water in itself.

Third, even with ambitious targets for growth of crop yield on existing land (accompanied by greater rates of fertilizer application and more pollution) and perhaps with overly ambitious targets for development of rainfed agriculture, irrigated area will have to be substantially increased to meet the food demands of 2030. While water recycling and other conservation techniques may reduce the need for developing additional water supplies, it appears that these supplies must be increased by at least one-third.

In sum, combining irrigation demands with the increased urban-industrial demands, and assuming the utmost in water conservation, the world will need to develop about 60 percent more water by the year 2030, a rate of increase of about 1 percent per year. L'vovich and White (1990) reach roughly the same conclusion in terms of the consumptive use of water in the next century (fig. 7). In my judgment, the task of water development becomes especially urgent because of the need to preserve water supplies for environmental uses and to defend this use of water from encroachments by the other sectors. If water shortages in developing countries are permitted to cause conflicts between economic and environmental needs, the environment will lose. These conflicts can be lessened by further water development, supplemented by water conservation to the extent possible. Both in economics and in ecology, and for the same reasons, water is, indeed, destiny.

Literature Cited

Abu-Zeid, Mahmoud, and David Seckler, eds. 1992. *Roundtable on Egyptian water policy.* Arlington, Virginia: Winrock International.

Baskin, Yvonne. 1993. Ecologists put some life into models of a changing world. *Science* 259:1694-96.

Crosson, Pierre, and Jock R. Anderson. 1992. *Resources and global food prospects: Supply and demand for cereals to 2030.* World Bank Technical Paper no. 184. Washington, D.C.: The World Bank.

Ehrlich, Paul R., Anne H. Ehrlich, and John P. Holdren. 1977. *Ecoscience: Population, resources, environment.* 3rd ed. San Francisco: W. H. Freeman.

Fernald, Edward A., and Donald J. Patton, eds. 1984. *Water resources atlas of Florida.* Tallahassee: Florida State University Press.

Frederick, Kenneth D. 1993. *Balancing water demands with supplies: The role of management in a world of increasing scarcity*. World Bank Technical Paper no. 189. Washington, D.C.: The World Bank.

Glavnaia Geofizicheskaia Observatoriia. 1972. *Agroklimaticheskii atlas mira*. Moscow: Gidrometeoizdat.

Hargraves, George H., and Zohrab A. Samani. 1986. *World water for agriculture—precipitation management*. Washington, D.C.: Agency for International Development.

Howe, Charles W., and W. Ashley Ahrens. 1988. Water resources of the upper Colorado River basin: Problems and policy alternatives. In *Water and arid lands of the western United States*, eds. Mohamed T. el-Ashry and Diana C. Gibbons, 169-232. Cambridge: Cambridge University Press.

Kasting, James F. 1993. Earth's early atmosphere. *Science* 259:920-26.

Keller, Jack. 1992. *Implications of improving agricultural water use efficiency on Egypt's water and salinity balances*. Center for Economic Policy Studies Discussion Paper no. 6. Arlington, Virginia: Winrock International.

Keller, J., N. S. Peabody III, David Seckler, and Dennis Wichelns. 1992. *Water policy innovations in California: Water resource management in a closing water system*. Center for Economic Policy Studies Discussion Paper. no. 2. Arlington, Virginia: Winrock International.

Keynes, John Maynard. 1935. *The general theory of employment, interest, and money*. New York: Harcourt, Brace, and World.

L'vovich, Mark I., and Gilbert F. White. 1990. Use and transformation of terrestrial water systems. In *The earth as transformed by human action*, ed. B. L. Turner, II, 235-52. New York: Cambridge University Press.

Lydolph, Paul E. 1985. *Weather and climate*. Totowa, New Jersey: Rowman and Allanheld.

Martin, Ruth E., ed. 1965. *Encyclopaedia Britannica international atlas*. Chicago: Encyclopaedia Britannica.

Moore, Deborah, and David Seckler, eds. 1993. *Water scarcity in developing countries: Reconciling development and environmental protection*. Arlington, Virginia: Winrock International.

Postel, Sandra. 1992. *Last oasis: Facing water scarcity*. New York: W. W. Norton.

Rensberger, Boyce. 1993. "Greenhouse effect" seems benign so far. *Washington Post*. 1 June.

Seckler, David. 1992. *Irrigation policy, management, and monitoring in developing countries*. Center for Economic Policy Studies Discussion Paper no. 4. Arlington, Virginia: Winrock International.

Seckler, David. 1993. World grain consumption and production: 1961-2030. Arlington, Virginia: Winrock International. Typescript.

Seckler, David, and Deep Joshi. 1982. Sukhomajri: Water management in India. *The Bulletin of the Atomic Scientists* 38(3): 26-30.

Snead, Rodman E. 1980. *World atlas of geomorphic features*. New York: Van Nostrand Reinhold.

Thornwaite, C. W. 1963. *Average climatic water balance data of the continents: Part II, Asia*. Technical Report no. 2. Centerton, New Jersey: National Science Foundation.

Unesco. 1978. *World water balance and water resources of the earth*. Paris.

USDA (U.S. Department of Agriculture). 1941. *Climate and man.* Yearbook of Agriculture. Washington, D.C.

Wade, Robert. n.d. *Inside strong and weak states*. Forthcoming.

Wittfogel, Karl August. 1957. *Oriental despotism: A comparative study of total power.* New Haven: Yale University Press.

World Resources Institute. 1992. *World resources 1992-1993.* New York: Oxford University Press.

Yudleman, Montague. 1993. Draft report on water resources to the International Irrigation Management Institute. Typescript.

Panel

E. WALTER COWARD, JR.
Ford Foundation

This paper reflects an important strength of Winrock. That is that it takes a topic such as water resources and deals with it by cutting across and integrating issues and experiences in the United States with experiences worldwide and in the developing countries in particular. My own view of Winrock is that this is a very powerful strength and capacity that I hope in the future we can find more ways to link with and to build on.

I want to talk about four points: (1) the progress that we've made in the water resource and irrigation area, (2) equity issues, (3) institutional weaknesses, and (4) what I see as conceptual limitations.

On the first point, what we have been able to accomplish in thinking about water resources and irrigation, I think we've made some strides. For people who work in and think about this area, the topic has now gone beyond engineering. That does not mean that engineering is no longer relevant. But we're much more concerned with a socio-technical way of thinking about irrigation: irrigation institutions, social organizations, matters of pricing and economics. When we think about water resources and irrigation now, there's much more attention to local people. Water-user groups have become an important factor in water and irrigation. In addition, there have been some important institutional developments. The creation of an international center, the International Irrigation Management Institute, which focuses on the topics we're talking about is very important progress. Also there is now much more concern about how irrigation systems perform. It's not just a matter of how to build more systems: It's very much a matter of how to maintain as well as increase and enhance the performance of those irrigation facilities that are already there.

A second point I want to make is that this sector of agriculture and environment is replete with equity issues. The fundamental problem in any irrigation system is the equity between those who are at the head or the tail of the system and how to adequately distribute water to them. There are also the significant questions at the national or regional policy level that have to do with equity issues between irrigated areas and nonirrigated areas. And of course, we are increasingly seeing equity questions arising as many people look to the agricultural sector as the source for water that they want to use for other purposes. If you ask how to meet the demands that many have for so-called environmental uses of water, most people have the notion that it will come from the agricultural sector. That's not a bad idea because that's where most of the water is now being used. But it does raise important equity questions. We're involved in some work in northern New Mexico where Hispanic communities have been using water for several centuries under rights originally provided to them by the Spanish king. Folks are coming in who want to build ski lodges and have other interesting ways of using the water, and there are difficult problems about how you respect water rights of people while also trying to accommodate progress and improvements into the future.

I would also say, in this equity arena, two other things. One is that we're still trying to figure out whether or not there are important gender dimensions to this. Many people have raised the question about women and water. It's a topic worth pursuing, but not one that I feel we have clear answers to at this point. And, finally, I would remind you that one cost of much water development around the world and over time has been that poor people have been marginalized by our so-called water resource development projects. This extends to everything from the problems that you get into in resettling people from reservoir areas through the problems of who will in fact get to use the water that is made available through the new projects.

The third point is that this is an area in which there are severe institutional weaknesses. And until we can get a handle on resolving some of these institutional problems, we're going to continue to struggle in this area. If you think about how important irrigated agriculture is to the world's food production, it's amazing how few centers of excellence there are with regard to irrigation. It's wonderful to have IIMI but, obviously, the whole world can't expect IIMI to solve all the problems of irrigated agriculture.

Some years ago when I worked in Indonesia, it was amazing to me that, important as irrigated agriculture is to the economy of Indonesia, there were essentially no centers of excellence and hardly any prominent scientists in Indonesia who worked on this topic. One of the great problems that arose was that Indonesian policy makers were continually subject to whatever was the newest idea about irrigation that foreign consultants brought to Indonesia. There was no capacity on the Indonesian side to say, "Now, wait a minute, we tried that" or "That is not a good idea for this place" or "We have different strategies or ideas in mind."

It's curious, too, that there is only modest NGO involvement in the water area. You cannot find NGOs that worry about water in quite the same way that you can find ones that worry about tropical forests or other environmental problems. In the Ford Foundation, we've been trying to promote some fledgling attempts to get the NGO community more focused and more motivated about the water sector. But at this moment, the NGO activity is relatively modest in this sector.

Also many of the hydraulic institutions that we have around the world don't fit very well with the hydraulic realities that David Seckler was talking about. If you look around the world, we have few institutions whose boundaries coincide with watersheds. That's not the way we put together our local governments or administrative structures. Similarly, most rivers cut across many jurisdictions and institutional arrangements. So we don't have a good fit between our institutions and the hydraulic regimes with which we're trying to deal.

Additionally, I would point out that irrigation bureaucracies themselves have tended to be weak organizations. While there is growing interest in the performance of irrigation systems, the ability to measure performance, to assess it, to monitor it on the basis of irrigation bureaucracies is poor indeed. I've always been astounded that we know a lot more about how the New York Yankees perform than we know about how irrigation systems in Pakistan perform. We don't have the kinds of records that tell us how well we're doing or not doing in this particular area.

The fourth point is that I want to underscore my perception that our ability to advance understanding of water resources and to work with them is hampered conceptually. Some of the problems that we are seeing with our models of economic development spill over to our ways of thinking about water resource development. Many of the concepts that are part of our fundamental ways of thinking about economic development are the same concepts that we use in assessing water resource projects and thinking about what we ought to move ahead with—deciding whether or not, for example, it's more important to maintain a salmon fishery or to irrigate rice lands in northern California. So, unless we're able to advance our conceptual understanding on that line, I think we will continue to have problems thinking about water resources.

N. S. PEABODY III
Winrock International

I've been working with David Seckler on the notion of the multiplier effect and the attempt to look at basins and use and re-use of water. This perspective is important for talking about the future because it integrates many concerns that are often neglected. What I would like to do is carry on some of the implications of using this perspective.

One is that it clearly identifies the interdependence between water users at various points. It implicitly argues for an integrated planning perspective, and an obvious unit is the water basin. Users are interdependent in terms of the quality and the amount of water that's available and used and the methods of water management and allocation. What somebody else wastes, you may lose. If the quality of the water you receive is good, you can use it well; if it's not, it's a burden, and that burden becomes magnified downstream.

Efficiency, too, increasingly is a public issue. How people use the water within the system makes a big difference to all users. If you look at this from a basin perspective, you may say that if water is wasted at the top, that's all right because it's re-used down at the bottom level when it returns to the stream. But that's not exactly true. It depends on where it goes when it's used at one point and where it is used at another so that, unfortunately, the broad model that focuses on an integrated kind of approach has limitations—if you don't look at the details of how the system operates internally, you can miss important considerations and draw false conclusions.

The public issue, in terms of efficiency, has to do with equity. Quality is another issue. And a third is the ability to increase the types of use of water by reallocation. Often the available water in a basin is largely captured by existing users, and methods to open the process to others are not simple.

So, getting back to efficiency, the way you focus on the use at one point to another depends on the conditions. If you are using too much water in an area that has a saline aquifer (for example, parts of Pakistan are problematic because in certain places the aquifer is saline, in other places it's sweet), that is a real waste of the resource. Water that infiltrates is lost if it goes to a point where it has no further use, like on the western side of the Central Valley in California. The only way of recapturing it and increasing the water multiplier is to clean the water—to remove the salts. On the other hand, if you waste water on the eastern side of the Central Valley, it percolates into an aquifer that people can draw from and, in fact, they actively use the aquifer for storing water. People who are using water first are encouraged to use as much as possible or at least to get as much as possible and have it infiltrate for storage.

Another point that emerges from the closed system model—and the thing that's rarely done—is the need to look at surface and ground water as a system. There's much talk about conjunctive use, but few good examples of it.

Over time, the ability to move water to different places for different uses is going to become increasingly important. Few places have "plumbing" like California, but others may develop over time.

Some of these points get back to the issue of the institutional framework in which water is managed. As Walt Coward said, water institutions are not very strong. In fact, if you look at the broad perspective of water in terms of its environmental impact, as well as economic impact for sectors other than agriculture, you see that the institutions that deal with water are fragmented and sectoral. If you're concerned with improving or maintaining water quality, there is no obvious institution to focus on in most societies. There are few institutions that have a mandate to manage water quality. Some do a little, but they're not coordinated for the most part. There is no mutual accountability.

Also, most of the institutions are overextended. Some irrigation institutions may have long histories, but their real lives have been the last 20 or 30 years when they planned irrigation investments or constructed irrigation systems. So they are huge and relatively new bureaucracies. Anything they knew about judicious management was lost during the construction period. At the same time, there is now a strong push to have these institutions do more than construction. They have to focus on operation and maintenance, and various other things, which because of the present incentive systems, are difficult for them to take seriously, and consequently they don't do them very seriously.

Another thing to note is that the institutions historically have ignored water quality issues and the impact of their work on the environment. This is true in the United States as well as elsewhere. We're learning, particularly in California, that the attempt to redress such oversight is not a simple process. One of the biggest lessons that is emerging is that the more flexible the system is and the more stakeholders participate in the planning and investment process, the more likely you are to end up with a system in which the water is actually well used. But that process is painful, as California found during 15 years of conflict between different forces—agricultural, urban, and environmental.

I agree that the conflicts are going to be growing internationally and nationally. We do not have institutions right now that can deal with these kinds of conflicts, but Winrock should position itself to help design them.

CATHERINE JEWSBURY
Winrock International

I've been involved with Eastern Europe and the Soviet Union, looking closely at immediate institutional questions that have to be resolved, and, in the process, taking stock of what the field of economics offers as strengths: where we economists need to extend our capabilities to address environmental issues analytically and in terms of policy advice. So, from my background, the main things I saw in David Seckler's paper that I think are interesting revolve around the equity issue. A fundamental shift of paradigm in the prices of water for environmental or for agricultural use, it seems to me, derives from the historical fact that we have gone through the development process, until recently, without taking the environmental uses of the natural resources in a sustainable way as part of our accounting process. Part of the equity issue is an intergenerational transfer that we're redressing, but we're also trying to redress inequity from point of view of the longer term future intergenerational transfers that might take place.

One useful thing that we can do in this competition for water use between environmental and agricultural uses is to address the issue of high value-added uses for water. That's something that clearly will have some equity effects that are worth looking at.

In the Soviet area, people are supersensitive to those equity effects. Often they ask me, as an environmental economist, about the links between the macroeconomic and the microeconomic. They are strict in their judgments about the microeconomic focus Americans have, which they feel lacks a macroeconomic linkage that is important for their situation. They frequently ask me about welfare. In their economic model, coming from the Marxist system, this is constantly on their minds. They don't know how to analyze it when they turn to a market system. We haven't given them the tools yet.

The human resources question and training needs that we've talked about point to a major way to address a lack of familiarity with market economics in general; but lack of understanding about how a market system deals with equity issues can also be addressed through institution-building activities and helping policy advisers develop the capability to integrate resource use issues into the larger policy making framework.

When I look at water questions, the NGOs in particular fit in an interesting way. When we consider the NGOs and the role that they are playing in these societies that are in transition, one difference between them and NGOs in Asia and Africa is that they are heavily involved in the democratization process and trying to create a role for themselves in adversarial positions, or at least in a testing situation, relative to the government and the power structure within political and academic institutions.

When we examine this particular area of water resource pricing, it seems to me that all of our experience elsewhere can be brought to bear, and we're looking at the conjunction of the four of the pillars of Winrock that Bob Havener listed, which are institutional, human resources, technical, and policy advice. It seems to me that the field of economics can touch very well on three of the four.

Discussion

Northrop: I was struck by David Seckler's characterization that environmental interests will be a growing user of water resources. Are there some examples?

Seckler: In the closing water-cycle model, with the exception of certain circumstances like highly saline areas—where if you use too much water you really do lose it because you get so much salt that nobody can use it again—what you really concentrate on is the consumptive use of water in each use, that is, how much water gets evaporated from the system as it's used. For example, 28 percent of all the water withdrawn for use in the United States is used to produce power, mainly in steam electrical generation and hydroelectric facilities. That's an amazing figure. But when cooled steam is released, what happens? It flashes off, it condenses, it falls immediately back into the system, and the real consumptive use of water from a boiler is virtually nothing. If you put that waste steam through a condenser, you just recycle the water perpetually. You don't consumptively use the water. That's the key idea. Similarly in hydroelectric plants hardly any of the water is consumptively used, that is, lost through evaporation.

If you take all of the water that urban and industrial users withdraw, probably 20 percent is consumptively used, that is, is evaporated away. The remaining 80 percent goes back into the stream flow somewhere and is re-used. In agriculture, the figure is something on the order of 50 percent, depending on the efficiency of the irrigation system.

When you get into the environmental uses of water, this is something the environmental community really hasn't understood. That is, they are in one of the most highly consumptive water-use sectors we have. Take wetlands, for example. Wetlands will be covered less than a meter deep and sometimes just a few centimeters deep over vast tracts of land. What that means is that most of the water is exposed at the surface, and wind and the heat of the sun evaporate it away. In a wetlands, I think you're running a consumptive use of water on an annual basis of around 100 percent.

When you calculate the economics per unit of consumptively used of water, I think you would find that, unfortunately, the environmental uses are generally going to come out at the low end—that per-cubic meter of water evaporated away, the economic value of water in the environment is less than in almost anything else. That's why I'm concerned about environmentalists going around getting economists to side up with them, because they haven't really figured this out yet. When they start saying, "Let's allocate on the basis of economic value," they're going to lose all their water. That's my opinion.

Goodwin: I want to argue with your definition of consumptive use because, in fact, the water that's evaporated gets recycled. It comes right back down. It would be bad if we got into a situation where no water was getting evaporated. We wouldn't have water. So, you're talking as if consumptive use by the environmentalists is a bad thing. It doesn't sound like altogether a bad thing. The only place where water is being really wasted, as you pointed out earlier, is where it has been mixed with something that makes it nonuseful and then held at that place. It's those aquifers or ponds or whatever that have bad stuff in them that are the real consumptive uses.

Seckler: That is both true and not true. It's true that on the globe, the water can't get away. It has been here for 5 billion years, it will be here another 5 billion. So, you don't have to worry about water in that sense. The consumptive use is important because most of it goes out of the region and back into the sea. So if you have a high consumptive use of water then what you're really doing is moving that water back into the sea—granted that it will be evaporated and come back to us again—but it may come back in outer Mongolia, not where we want it. If you're thinking at a less than global level, you have to worry about consumptive use in a region.

Peabody: In California, when engineers talk about using water that's available, their usual assumption is that any water that they don't pick up—that ends up in the sea or a salt sink—is wasted water. So their objective has been to use as much of the water as possible at any point—to capture it and manage it using their best engineering skills. Because of the Endangered Species Act, a minimum flow now must be maintained in certain rivers. Previously, if there was no water flowing at certain times of the year, it was no problem to the engineers. That means they have had to re-factor the allocation of the water, effectively reducing the amount of any water that they've impounded because some of it must be released at times to maintain minimum flows. The consequences of maintaining minimum flows is important, and certainly the timing is important because it can have a real impact on the amount of water that's actually available for use in agriculture.

Northrop: I wonder about characterizing that as an increase in use. I would see it as a giving back to a source that has been lending it, in a way, for a long time. So, to characterize it as a net environmental use of the resource of water seems misleading.

Grove: I would like to reinforce that. We have to look at the alternatives. There are certainly alternatives other than keeping birds and bunnies alive or growing food. Indeed, if I do an analogy, not with withdrawing irrigation water but with the attitude that has occurred in North Carolina: Agriculture is king and let the stuff go in the creek. But the second and largest productive estuary on the East Coast no longer supports a number of the fisheries that were there. We can't harvest shellfish from any waters of the state anymore. So, we must keep in mind that there are some downstream, off-site kinds of economic values, and perhaps some other values, when we think about these environmental consumptive uses. They aren't just luxuries.

Coward: There is a disadvantage in setting up a dichotomy between agriculture and environment. When you put it that way, environment sounds like a passive thing that doesn't have any particular economic values attached to it. Or if there are, they are off in the future and we don't know how to count them, etc. I think we need to talk about agriculture versus fishing, fisheries' uses of water, and other economic things, because that will clarify the choices we're facing. I don't think it comes out that way when you say, agriculture versus the environment because it sounds like the environment is something nebulous.

Seckler: Let me try to clarify. Much of the environmental community, particularly in California, has been going around saying, "If we allocated water on the basis of its true economic value, then we would have much more water for the environment because it would cause people to save water, and then we would have more to protect the birds and everything else." As an economist, my advice to them is, don't push that very hard because people will ask what is the value of water in environmental uses? My advice is don't get into that argument because I don't think economics pertains to a lot of environmental values.

If we want to protect wildlife, do it on noneconomic grounds. Don't get sucked in by the economists. You're going to lose the battle. The economic value of water is going to be higher in other uses because now we understand that environmental use is a major water consumer. Environmentalists haven't quite gathered that they are really causing enormous quantities of water to evaporate. I'm in favor of that and I want to keep it up. I don't want them to get that trapped by economic value judgments.

Harwood: Isn't it necessary to begin to put a market value on water according to quality and begin to devalue it as you load it with the bad stuff, whether they be nitrates or the various toxins, or whatever? Then, as you flow it through a particular use, you not only have the consumption but the devaluing of the upflow.

Seckler: Yes. The water multiplier idea stresses that you have to keep your water clean as you're going through these iterations, otherwise you effectively lose the quantity of water you would have. It stops the multiplier. In Egypt the research emphasis in the future is going to be on how to keep the salinity and other pollutants out of the water, so you can maximize the water available for whatever uses you want to put it to.

Berg: Wetlands are not the only dimension of environmental quality or degradation of the water. Suppose you think of poor health as an environmental consequence of the way water quality is degraded. Do you have an opinion on what the economic payoffs to improvements in quality of that sort might be?

Seckler: I don't get too excited about the economic payoff in the health dimension because I don't think we can estimate it, and I think we ought to have safe drinking water for people as kind of an absolute. I'm quite sure that you would find very high returns to safe drinking water. You would find that it justifies almost any investment you want to make in that field. That doesn't mean that you have to keep all the water supplies safe, but at least the drinking side. You have to be particularly careful about the heavy metals that people are dumping into water supplies. The Ganges is full of heavy metals like cadmium and arsenic from the tanneries.

Berg: I have a question that's based on perfect ignorance. If the Secretary of State wanted to gear up discussions on one of these areas again, like the Middle East, and said, "Where do I go for expertise on this?" are there people or institutions that have actually been working on this or thinking about solutions?

Seckler: There is no institution that's even remotely capable of doing that, but there's a good network of people around the world that have thought about these problems that can quickly be assembled. In fact my own philosophy is that networks beat institutions hands down, and that in a modern age of communication, we shouldn't have so many institutions doing this kind of thing. We should just be networking back and forth.

Coward: I see it slightly differently. For example, there is a strong team based at the University of Pennsylvania with Tom Naff and others who have been working on this topic for a long time. They were quite successful, for example, last year in bringing both Palestinians and Israelis together in Geneva to talk about these topics. I think there is some strength there. There are, at the University of New Mexico and the University of Arizona, groups that deal specifically with transboundary water problems. So there is, I think, a little more institutional capacity than, perhaps, David was suggesting.

Berg: Are some viable solutions on the Middle East out there that the network has already identified?

Seckler: There are places in the Middle East where there are absolutely no solutions to the problem. Literally speaking, Egypt, as best we can estimate, can perhaps generate 10 percent more water without going to war with somebody, let's put it that way. They only have about 10 percent more water and their population is growing at over 2 percent a year. They are increasingly importing food, and they will wind up importing all their additional food requirements one way or the other.

The Middle East has probably the worst water crisis on earth. Libya is worse than any other country because all their water is coming from pumping fossilized aquifers. They have almost no precipitation or rivers; they're just mining water. So they're just flat going to run out of water.

Peterson: From my experience in Yemen, there are a lot of water savings just in terms of management of water that can be improved in the Middle East. Water management doesn't seem to be one of the strong points in the Middle East right now. With the exception of perhaps Israel, there aren't a lot of water conservation measures going on.

If you want to invest, would you invest in irrigation delivery or would you invest in drainage?

Seckler: There are places where drainage is extremely important but that certainly isn't as high on my list of priorities as irrigation. I would invest in large diversions and storage, more water-supply building facilities.

Peterson: Would you address the salinity problem caused by poor drainage and irrigation as a soils issue or would you address it as an irrigation issue?

Seckler: Soils. The only place that salt comes from is the soil.

Peterson: You engineers are all alike.

Seckler: There are saline areas where they shouldn't be irrigating, but they have to, like Pakistan. Then there are places where they shouldn't have invested in irrigation because the soil is too salty. But there's nothing really to do about it. Probably the Imperial Valley should have never been irrigated for the same reason.

Blake: I would like to, perhaps, throw a bomb here. What is probably implicit in what has been said, but not explicit enough, is the urgency of the problem that we face, that is, putting in clear terms the key role irrigation has to play in providing enough food fast enough to meet the problems of the world. It is being underestimated in every way—institutionally, in investments, and in research. It's almost as if people didn't realize that within 20 years we have to increase food production by 50 percent and that irrigated agriculture will have to provide most of this increase. There are big policy implications to that. We have no time to get new institutions in place, to get new investment policies in place, to stimulate the major investment sources like the World Bank and UNDP to make a bigger contribution.

I also would urge that we not overlook the fact that, at this point, we don't know how farmers, particularly small farmers, can radically increase production with deteriorating soil and less water and, do so within a decade. And this must be done on less land, if you consider the rapid deterioration of the quality of irrigated soils all over the world.

All this means that we must pay attention to the inadequacy of irrigation research—the inadequacy both in quantity and quality of what is being done. What IIMI is doing is important, but it's a tiny program and there's nothing backing it up in national systems in most countries.

David Seckler recently pointed out the urgent need for good conceptual, strategic, across-the-board planning by countries that are faced with this problem. Nobody is getting to them in a holistic way to look at the costs of inaction, and what they must do. If there's one flag that should go up from this conference, it would be to say, "Attention, everybody, here's a problem that has to move now."

Coward: David Seckler said that the food needs of the future are going to rise because of increasing need for foodgrains for animals and that doesn't require irrigation. So how does that stack up with what Bob is saying here?

Seckler: I can't say enough for the study by Crosson and Anderson. It puts a new twist on the future of agriculture. So much of it will be in the feedgrain crops. I agree with Ambassador Blake that whatever the future looks like for 50 years out, we know we're going to need a lot more irrigation. We must remember that if we're going for rainfed agriculture in Africa and Latin America, not only may the environmental consequences be severe, but the sheer cost of putting in transportation infrastructure into those rainfed areas will be fantastic. So I would agree that the decline in irrigation development—old-fashioned irrigation: just shove the water out, get more water out on that land—is probably the most dangerous trend in the whole world food situation.

Crosson: In the work that I did for the World Bank, it appeared that, according to bank estimates, there might be potential for increasing the amount of currently irrigated land by as much as 50 percent. But those estimates really reflect judgments about how much land there is around the world. They considered the physical, climatic, soil, and terrain features necessary for irrigation, but they left out of account the economic cost of bringing that land in, as well as the environment cost. It was on the basis of that kind of argument that it seemed to me that the potential that could be realized at economic and environmental costs consistent with sustainability would be a lot less than 50 percent, without knowing how much less that might be.

One of the things I then thought about was, what are the opportunities for using present water supplies in irrigation more efficiently? Then I ran up against the same argument that David Seckler made, that if you look on a system-wide basis, the efficiency is not all that bad, maybe 70 or 80 percent, implying, therefore, that there is not all that much opportunity for getting more out of increased efficiency.

If Bob Blake is right that we will have to have big increases in irrigated land to meet future demands for food, that the critical question is, how much potential actually is out there that could be realized under sustainable conditions?

Seckler: For Asia, there's room for increasing the irrigated area maybe 25 percent. But it's alarming that the growth of urbanization in Asia, which occurs primarily on irrigated land because that's where the cities are, could amount to a third of the irrigated area of Asia by 2030. So even if you invest in new irrigated area, you may not gain in terms of total irrigation.

It's a disgrace that when we have projects like India's Sadar/Sarovar project, in which you could bring in 2 million hectares of new irrigated land in an arid, poverty-stricken area, we stop because people are confused over the environmental and resettlement issues, which could be managed. These are the kind of things that we just can't afford to do. India will go ahead and do the project, and they'll do a worse job, especially with regard to the environmental concerns because the World Bank has backed out. We need to support some of these big projects.

Without knowing much about it, the Three Gorges Dam in China, it seems to me, is essential to China's development.

Sub-Saharan Africa has a potential for about a 25-million hectares of irrigation development. About 4 million of that is already developed. So you have another 20 million hectares you can put in. That's only a third of what India now has. This area should be developed, but then sub-Saharan Africa is over with.

We have to find ways to improve rainfed agriculture, but we all recognize that people have been working on rainfed for 100 years and the record of success in arid areas has been depressing.

Havener: Most people assume that new irrigated land will cost a lot of money to develop and that increases in rainfed agriculture will not cost a lot of money. In my view, that's not reality. Significant increases in rainfed agriculture will require immense investments. David Seckler talked about infrastructure, meaning roads and those sort of things. But the technology needed to bring about increased productivity in rainfed areas is going to be substantial because, after all, it will be virtually impossible to break the connection between water availability and increased productivity. If we're going to do that under rainfed conditions, it means we're going to be draining and we're going to ditching and we're going to be damming and we're going to be engaged in water-spreading and we're going to be putting in physical structures and land leveling and terracing and so forth. Without these investments we are not going to get substantial increases from rainfed agriculture. It will be a different infrastructure but there's no reason to assume that it will cost less per calorie to get them.

Peterson: Are you saying we can't look for significant increases in productivity unless we have irrigation or some way to manage the moisture regime?

Havener: I'm saying moisture regime management, which may or may not involve irrigation, is the important variable. I believe we're going to think about irrigation much more broadly than we have for some time. Sure, life-saving irrigation in areas of 500 millimeters of precipitation has a high rate of return, provided you have it where you need it, when you need it, and as you need it. I'm also talking about terracing and water collection and spreading and many other physical infrastructures that may not be conventional irrigation but, nevertheless, will provide moisture within the soil profile for a growing crop.

Maner: In Guatemala, we do not depend only on irrigation. We've been working on rainfed agriculture but supplementing it with irrigation from small watersheds. We're not talking about large, flat irrigated areas, which is what World Bank and other people tend to think about. We don't have to have large-scale irrigation systems. In watersheds that have water available, farmers in Guatemala have taken small plastic pipes and moved water down the watershed. Irrigating by hand with tubes has increased their income and food production tremendously. I think we need to think about other systems that might be utilized besides the large-scale irrigation systems.

Seckler: While I spent a good part of my life doing little irrigation systems, and I love them dearly, I think we must face the fact that they are very management intensive. They are very hard to apply to large areas of land. Also, the costs are high. You get enormous economies of size in irrigation systems. The cost per unit of irrigated area or per kilogram of product in these small systems is usually two or three times as much as it is in a large system. Even though it's important in specific circumstances and from an equity point of view, they really are not alternatives to the large-scale storage diversion and projects that we need.

Maner: The many irrigation systems we've put in have been very inexpensive. It really amounted to putting a hose a kilometer up the creek and allowing gravity to bring the water down for irrigation. Little infrastructure has gone into development and use of these small systems, and they've been effective.

Coward: I don't think that, historically, there's any basis for saying small-scale irrigation can't make a big difference in a particular country. In a number of countries around the world, you find that large portions of the irrigation sector are currently served by small systems. In the Philippines, about half of the irrigated area is in small-scale systems. In Indonesia, it's about 40 percent. But it's an equally large proportion in places like Nigeria and Peru. So, there's quite a geographical spread.

I'm not satisfied with the notion that small-scale costs more than large-scale. It seems to me that if you add in the costs of the bureaucracies that are required to operate and maintain the large systems, you end up with very different results. Most of the small systems are going to be maintained by local people. We also know that many small systems are, in fact, pump irrigation, lifting ground water. I think most people have felt that where people have to pay in such a direct way for lifting, it's a very efficient way of irrigating.

Kaul: I thought we had a statement that in semi-arid areas, the types of crops grown are the ones that are most suitable for livestock agriculture and that's where the future of the dry areas is. I disagree. Seventy percent of the poorest of the poor live in the semi-arid areas. These are the very areas where there is maximum diversity in crops, which have not yet been fully explored. The local populations sustain themselves on these crops and that is where the potential is. Technology has not been developed because it's a difficult agriculture compared with irrigated agriculture. We have not yet spent much on understanding it and perfecting it and bringing it to use. To say that in this the area, the future lies in producing livestock because sorghum and perhaps maize would be more suitable for livestock-raising would be a serious mistake.

Seckler: I agree that we have to think not just about irrigation but about water regimes. I'm talking about those fairly wet areas in central Latin America and the millet-sorghum belt that extends across the northern part of sub-Saharan Africa. These are human foods and they're also livestock feeds. So, it's suitable in that sense. In India I tend to agree. It's too dry in most of those places even for those crops to do anything.

Paarlberg: If there's a sensible place to pursue rainfed feedgrain production, it would be either in the southern temperate zones, Argentina and southern Africa, or in the northern temperate zones, in Illinois and Iowa. Commercial trade can then move these feedgrains to their final point of consumption. For countries in the tropics, feedgrain shortages usually reflect high income, which is usually an outgrowth of rapid industrial development, which in turn generates the foreign exchange that's needed to import the feed from the temperate regions that are better suited to produce it. I would propose, as a sustainable solution, international commerce rather than an overweening effort to boost rainfed feedgrain production in tropical countries.

Crosson: In the piece I did for the World Bank, I emphasized that any country should keep in mind that there's a robust international trading system in agricultural commodities. Countries that are beginning to run up against limits to what they can sustain in agricultural production ought to keep that option open.

I think it would operate for most countries at the margin. For many countries whose income is growing fast enough to generate increased demand for grain and animal products, with the consequent impacts on grain production, the only way that income can be growing that fast, it seems to me, is through progress in agriculture. So there's a kind of circularity here on the basis of that kind of thing, as well as political constraints that governments face about how far can they go toward substituting imports for domestic production.

I just made a judgment that most of the increase in demand in developing countries would have to be met from production in developing countries. The opportunities for using the trading systems and bringing in imports certainly should be used, but that there are limits on how far—both from an economic standpoint as well as from a political standpoint—governments would be willing to go in that direction.

Raun: In Latin America and in Asia, the so-called subhumid areas, say, in the range of 1,000 to 1,500 millimeters rainfall, are relatively developed. In contrast, in sub-Saharan Africa that is not so. One reason is poor infrastructure. But another is the presence of tsetse fly and trypanosomiasis. They are still a barrier to the development of these areas, but the situation is changing not only because of advances on the technology side both as related to the vector and the disease but also because of human population pressures. Nigeria is a prime example. It's pushing the forest back. As this is done, as people come in with their crop-livestock systems, they remove the tsetse habitat and, in turn, reduce the incidence of trypanosomiasis. Sub-Saharan Africa is finally going to get to the point that it can be self-reliant if not self-sufficient. The only area that has major potential is this subhumid band. To exploit that and to develop it, it will take an organized, integrated effort. But there is potential there.

4
Poor People, Resources, and the Environment

Introduction

LOWELL HARDIN
Purdue University

As has already been made clear in this symposium, means of achieving a better tomorrow for the world's poor people and sustaining the productivity of our natural resources cannot be considered separately one from the other. While the proportion of the world population that is in poverty may decline, the absolute numbers of the poor will likely rise. In fact, it is estimated that currently in excess of 500 million of the world's 5.5 billion people do not have access to enough food to meet energy and protein requirements for a healthy, productive life.

Rapid population growth, grinding poverty, low agricultural productivity, uncertain or nonexistent tenure rights, and centralized, urban-biased governments are the handmaidens of natural resource degradation in developing countries. To the rural poor attempting to cope under such circumstances, concerns about long-term sustainability rank a far-distant second to meeting today's basic survival needs. The result is often unsustainable if not irreversible exploitation of fragile lands.

One response to these mounting threats to the natural resource base is to turn, as we have done traditionally, to yield-enhancing agricultural technology. By increasing the yields in the high-potential areas, pressure on the more fragile land and water resources can be reduced. Fair enough—if we know how. Current analysis of data from highly intensive rice monoculture and rice-wheat production systems in Asia raise cautionary flags, however. In some settings, more and more inputs appear to be required to achieve the same level of output. Scientists have yet to determine precisely why this phenomenon is occurring. It suggests, however, that even now some of our most productive farming systems may not be sustainable. And it is on these systems that we rely to produce the surpluses that help keep agricultural product prices down and provide the food for ever-mounting urban populations.

By no means are poor people the only driving force contributing to environmental degradation. Many of the prime causes are the result of externalities imposed by the "rules of the game" under which societies produce and consume. While not ignoring the impact of the high levels of resources employed in meeting consumption demands of the affluent or the negative environmental impact of misguided public policies, I emphasize the need for the donor world to continue the battle to reduce poverty. True, many efforts have met with little success. True, with the demise of the communist threat, one of the West's principal motivations for supporting anti-poverty programs has disappeared. True, there is much legitimate disenchantment about collaborating with unresponsive central governments in some of the developing countries. True, disappointment with the impact that overall economic growth is perceived to have on the well-being of the poor is causing development assistance managers to turn increasingly to applied community-level projects in collaboration with nongovernmental organizations. Nevertheless, on grounds of the donors' enlightened self interest, humanitarian concerns, and environmental protection, poverty reduction must remain a high-priority goal.

As we examine the interactions and linkages among such factors as rising population pressure, poverty, resource management, environmental quality, political power, and economic development, we who are especially concerned with rural development and resources need to be cognizant of what agriculture can and can not contribute. Increased nonfarm employment opportunities are a must, and these do not spring primarily from agriculture. This truth notwithstanding, the word agriculture seems to be dropping out of the vocabularies and budgets of development assistance agencies. This is occurring despite the fact that, during the 1980s, total food production in the developing countries rose 39 percent and per capita production went up about 13 percent. Even in Africa, total food production rose 33 percent although population increased even faster, resulting in a 2 percent decline on a per capita basis. The extraordinary performance of the agricultural sector is further demonstrated by the fact that, according to the World Bank, during the 1980s food and agricultural prices dropped by approximately 6.5 percent per year. In the absence of favorable price incentives, one must attribute much of this sterling performance to the development and use of cost-reducing technologies. Yet support for the research and education and investments in agriculture that help make such advances possible is diminishing rapidly. Is it perceived that agriculture has performed so well that it has worked itself out of a job? Certainly no one would hold that natural resources are managed in an optimal manner. We look to this symposium to help us better understand and address the reason for this short-sighted phenomenon.

Political Sources of Agricultural Resource Degradation in Poor Countries

ROBERT PAARLBERG
Wellesley College

We all know the familiar argument that poor countries supposedly abuse their rural environment because they are impatient to grow rich, and so they cannot afford to spend the limited funds that they have available to protect the environment for future generations. Poor farmers in these countries supposedly abuse their own soil and water resources because they are living from hand to mouth—their discount on the future is too high. They cannot afford to wait for trees to mature, so they do not adopt agroforestry. They cannot afford to wait for rangelands to recover, so they continue to overgraze. They cannot afford to wait for investments in terracing to pay off, so they continue to plow up hillsides.

Despite these arguments, poverty alone is a poor predictor of agricultural resource degradation, just as wealth alone is a poor predictor of resource protection. Poor farmers in developing countries do tend to abuse the resources in their environment, but not so much because they are financially disadvantaged. More often, it is because they are socially or politically disadvantaged. It is because they have weak control over the social or political terms of access to the natural resources in their environment. That control has too often been denied to them or taken from them by more powerful rural elites, such as landlords, or by their government, or by the powerful urban-based commercial interests, domestic and foreign, that work in close contact with their government.

The disadvantageous relationship of poor farmers to their own governments in much of the developing world is a point worth remembering. Here in the industrial world, democratic governments are often financially generous toward agriculture—some people would say too generous. Here in the industrial world, governments can be looked to, with some justification, as a means to solve rural environmental problems. In too many developing countries, unfortunately, there are what Pierre Crosson has politely referred to as "institutional deficiencies." Governments tend to be nondemocratic. They tend to be nonaccountable, especially to rural populations. They tend to be urban-based and urban-biased. For these reasons, they are quite often not the solution to rural environmental problems; they are more likely to be a source of those problems.

Poverty Alone Not The Cause

Why doesn't poverty alone explain agricultural resource degradation? Because there is plenty of evidence that poor people in desperately poor countries, when they enjoy adequate control over their environment, are capable of taking a long-term perspective. Some of the poorest farmers in the poorest countries of Africa cultivate slow-maturing perennial crops, under the right circumstances. Some of the poorest pastoralist communities have in the past managed their common rangelands effectively enough to prevent overgrazing. Some of the poorest hill farmers, for hundreds of years, constructed elaborate terracing systems and invested in the maintenance of those systems.

Second, poverty alone cannot be the explanation for resource degradation because some of the worst environmental damage being done in poor countries is being done not by poor subsistence farmers, but by relatively well-to-do commercial farmers. In Mexico, large-scale fruit and vegetable growers are overirrigating and jeopardizing farm-worker safety by spraying too many potentially dangerous chemicals. In Brazil, some of the most objectionable cattle ranching schemes in the Amazon are not being initiated by resource-poor farmers from the Northeast, but by millionaires from Sao Paulo.

Third, poverty alone cannot be the explanation because, as noted above, it is often governments themselves that initiate resource degradation. They do so when they arbitrarily nationalize forestlands and then sell off the timber for their own advantage or for the advantage of politically connected insiders, rather than for the advantage of rural dwellers. Governments do so when they appropriate otherwise well-managed common property resources in poor rural communities to operate land-titling schemes that primarily benefit the already landed or those that have political contacts with the government. Governments are the source of the problem when they displace environmentally sustainable fishing and farming activities by building hydroelectric dams to provide more subsidized energy to inefficient urban-based industries, rather than to the rural dwellers in the communities that are disrupted. They do so when they try to solve problems of rural landlessness by setting up colonization schemes in fragile, low-potential areas, like the interior of Brazil, the Yucatan of Mexico, or the outer islands of Indonesia. More generally, these urban-biased governments reduce the profitability and, hence, the likelihood of environmentally sustainable agriculture by underinvesting in agricultural research, by underinvesting in rural infrastructure, and by managing agricultural price policies and exchange rates to the disadvantage of producers of all tradable goods, but particularly agricultural commodities.

I should emphasize that it is not the poverty or the indebtedness of governments that makes them behave in this predatory fashion toward their own rural resource base or their own rural population. It is more likely to be their political nonaccountability to rural dwellers. After all, some of the worst abuse of rural resources has been done not by poor governments but by relatively affluent governments. Consider the mining of deep aquifers by oil-rich governments in Saudi Arabia and Libya or the damage done in the Aral Sea region by the former Soviet Union, which was not a poor government at the time. What these governments have in common is not a lack of resources, but instead a lack of accountability to ordinary people in rural areas.

Importance of Local Control

Why is government accountability to local communities so important to environmentally sustainable agriculture? First, even when governments (or outside donors) are well-intentioned, they usually have inadequate knowledge of how to protect the local resource base. Especially in the developing world, where farming environments are so difficult and so various from one river valley to the next, agro-environmental protection initiatives have to be highly site-specific. Their success depends upon intimate knowledge of highly localized conditions: the slope of the land, the build-up of pest populations, the seasonal availability of labor, tribal or caste divisions in the village or community, and locally appropriate roles for men and women. These are things that outsiders cannot possibly know as well as insiders, no matter how well-intentioned they are.

Second, we should admit that outsiders are not as likely as locals to be well-intentioned in the first place. It is local farmers and farm communities that will be most highly motivated to protect farming resources because it will be they, their children, and their grandchildren who will lose the most if that resource base is degraded. Outsiders, including governments, assistance agencies, and even NGOs like Winrock, just do not have the same stake in the long-term outcome. Outsiders can walk away from their mistakes more easily than insiders. To that extent, if outsiders decide to assume control over the resource-management problem, they will be more likely to make mistakes.

Political Weakness of Rural People

Finally, we should remember that it is naive, when we are talking about exploitable resources, to assume good intentions. This will especially be true in developing countries, where a majority of rural people will have difficulty protecting themselves because they may not come from a leading caste or tribe or because they may not even be literate. The temptation outsiders feel to exploit resources can become impossible to resist where standards of governmental accountability are not so well developed; where democracy, press freedom, and rule of law may not exist; and where financial betterment through respectable private commerce or enterprise is frequently not an option (because the private economy is so heavily regulated by the public sector). In countries such as these, many of the elites who serve in government, and many of those in the circles that influence government, are engaged in public life not exclusively to serve the country. They are engaged in the public life, in large part, to serve themselves, their families, their friends, their clan, their tribe, their region. National interests, and the interests of rural dwellers in particular, frequently get lost in the process.

These matters are not always discussed in public, especially among development assistance advocates who worry about flagging popular support for providing assistance to foreign governments. But in the developing countries themselves and, especially, among dispossessed and powerless rural dwellers, these harsh realities are anything but a secret. They are often a dominating fact of daily life.

How did it come to be this way? In many cases, the governing classes of today's developing countries are simply picking up where the European colonizers left off. During the colonial period, institutions or projects were designed to exploit rural resources for the benefit of outsiders, for the benefit of centralized institutions, and for the benefit of colonial administrators, with little regard for local communities. After independence, these predatory attitudes and institutions did not disappear. They largely passed into the hands of indigenous political elites who were mostly urban-based and who had weak knowledge about the daily lives of rural dwellers, just like their colonial predecessors.

Governments in many developing countries are not really strong enough to engage in consistently successful predatory actions in their own rural areas. They simply do not have the capacity in terms of size or degree of control. But governments in poor countries do not have to control the countryside in order to despoil it. They only have to be powerful enough to weaken or threaten the traditional institutions of local control. The local farmers themselves will then start doing all the damage. As Alan Durning has argued, nothing incites people to deplete forests, soils, or water supplies faster than fear that they will soon lose access to these supplies (Durning 1989). Simply because of the feelings of insecurity that poor farmers have, faced with the prospect of arbitrary dispossession, they will grab what they can. They will cut as many trees as they can, deplete as many soil nutrients as they can, and use up as much irrigation water as they can, before someone comes to take it away or someone comes to push them off to somewhere else.

In Ethiopia today, poor hill farmers are plowing up terraces in order to mine the soil nutrients for current crop production because they are unsure of what their new government is going to provide in the way of secure land tenure in the years ahead. Better to use what is there now while you have access to it than risk being pushed off, leaving it for someone else.

A Full Spectrum of Abuse

How can outsiders like USAID or the World Bank or Winrock respond to these difficult political sources of resource abuse in developing countries? I would argue that outside assistance agencies should recognize the problem for what it is. It is not just a technical problem, nor is it even a "policy" problem. We need technical change, and we need policy change, but we also need to encourage political change.

I am often frustrated by the apolitical nature of most technical debates about what sustainable agriculture should look like in the developing world. We argue endlessly about whether farmers in poor countries should try to protect the environment by moving ahead to modern science-based, high-external-input, intensive farming systems or by embracing low-external-input, traditional systems. This is an impassioned debate that usually pits overly technical agriculturalists and economists against overly romantic environmentalists. But it is usually an indeterminate debate because it abstracts away from politics.

In many developing countries, for example in Latin America, environmental damage is being done by farmers at both the high-input and the low-input ends of this scale, according to patterns that are politically determined. At the high-input end, wealthy commercial farmers frequently use their political power to monopolize the best irrigated lands and then to secure generous input subsidies from their government. They will then predictably abuse these lands through overirrigation or excessive chemical use. But at the same time, and often in the same country (sometimes in the same village or valley), poor low-input farmers—having been rendered landless or nearly landless because of their lack of political power or social status—will be using too few inputs rather than too many. They will be mining the soil, planting subsistence crops on sloping hillsides, and cutting into forest margins. They will be abusing the environment in a low-input fashion rather than a high-input fashion.

In countries where power, status, and access to good land are all so unequal, as in much of Latin America, parts of Asia, and even in parts of Africa, correcting resource management errors such as these is not just a question of providing better abstract technical advice. It may also require changing the political inequities that are compelling mismanagement in the first place.

At which end of this politically determined spectrum of technical mismanagement can we expect more progress in the future? I think that we can be mildly optimistic about reducing high-input agricultural resource abuse in the years ahead because it usually depends upon the extension of generous input subsidies to commercial farmers. These days, with many developing countries suffering from large debts and with many governments in developing countries under the discipline of structural adjustment, such subsidies are more difficult to afford. Indeed, they have already been cut back.

I am not as optimistic about what can be done quickly to reduce low-input abuse, which is engaged in by landless or powerless farmers in the developing world. To reduce this abuse, it would be necessary, first, to slow rapid rates of population growth. Second, it would be necessary to restore the control that ordinary farmers ought to have over the resources in their immediate environment. For the purpose of restoring local control, policy reforms (like structural adjustment) will not be enough. Structural adjustment is usually limited to changing price relationships, liberalizing some markets, maybe privatizing some state-owned industries, and getting exchange rates right. This is fine, and important, but it seldom contributes directly to the empowerment of rural communities. Structural adjustment is a process that is quite often negotiated between government ministry technocrats and specialists from the World Bank or USAID. It is a process that tends to be monopolized by highly centralized, technocratic institutions that are almost by definition out of touch with, and not responsive to, the site-specific, localized needs of those facing rural resource-degradation problems.

Mexico, for example is correctly viewed as a success story in structural adjustment. But Mexico's structural adjustment policies have done little to enhance the political power or the social status of poor farmers and farm workers. After half a decade of structural adjustment, Mexican farm workers still cannot legally organize. Their power situation vis-à-vis their own government is scarcely any better than it was before this important but overly centralized exercise got under way.

I should emphasize that when I talk about local control, I am talking about more than what is conventionally described as the spread of democracy. I am all for democratization, but it too falls short. The spread of democracy does not necessarily empower ordinary farmers in rural communities. There has been a remarkable wave of democratization across the world in the last half dozen years. It started in Latin America and spread to Central Europe and the former Soviet Union, and it has had an influence in Asia and even in some parts of Africa. But this has been a spread of democracy driven forward, in almost every case, by a newly powerful urban middle class, not by rural dwellers (landed, landless, or otherwise). The results have not yet added much to the power position of disadvantaged rural dwellers. In some places, as in Brazil or the Philippines, the embrace of constitutional democracy may have actually set back the power position of landless rural dwellers. Under the new constitutional systems in both of those countries, landed rural elites have used their legislative dominance to lock into law the existing distribution of rural land ownership. Remember, the most radical and successful land reforms that we have seen in the post-World War II period were not achieved by democracies. Instead, they were achieved by Japan while it was still under U.S. military occupation, by authoritarian or military regimes in South Korea and Taiwan, and by a revolutionary Marxist-Leninist regime in China.

Challenge for Outsiders

The challenge for outsider organizations, including Winrock, is to find ways to empower rural communities with or without structural adjustment policies and with or without a formal embrace of democracy in a capital city.

Often this is best done by avoiding governments, to some extent, either by working through nongovernmental organizations or by constructing direct people-to-people or community-to-community relationships. This can be done, for example, by helping to train otherwise politically or socially marginalized agricultural women in Africa; by organizing farmer-to-farmer programs in Africa or Asia or in the former Soviet Union that give rural agricultural people new information and the self-confidence they need to take better control of their future in fast-changing circumstances; by supporting rural microenterprise development in order to give rural communities constructive ways to build wealth outside of the public sector; or by launching on-farm seed management and development programs in Africa, built around the continuous local presence of well-informed, nongovernmental organizations.

All of these things, I am pleased to say, are currently a part of Winrock's program, and to that extent I think Winrock is responding to this challenge in an appropriate way. The environmental or the resource-protection payoff from projects such as these may at first seem minor, indirect, or even nonexistent. But, if my diagnosis is correct, helping rural communities and individual rural dwellers to gain greater control over their own fate is probably the most direct means to improve the environmental sustainability of their farming practices.

Literature Cited

Durning, Alan B. 1989. *Poverty and the environment: Reversing the downward spiral.* Worldwatch paper 92. Washington, D.C.: Worldwatch Institute.

Panel

SANDRA MILLER
Winrock International

There are real problems as we look at the issue of local participation. How do you encourage local participation? And, implicit in that, how will reallocation of power come about and what role should NGOs play? I believe that NGOs in general have something unique to bring to development and that sometimes we lose sight of their unique strengths.

I have spent my career working in nonprofit organizations and I think that too often we try to imitate public agencies, or universities, or private businesses. In doing so we lose the one thing that is our strength—flexibility. We cannot do what governments do in terms of mounting programs that require large outlays of funds. We cannot do what the private sector does in terms of bringing buyers and sellers together, creating markets, and reallocating products through those markets. What we can do is bring together groups of people and help them solve problems. In bringing groups of people together at the local level to solve problems, we come at the issue of reallocating power from another direction. If you address the issue of reallocating power directly, you will inevitably lose. If you address it from the position of solving a problem that people agree is a common problem, then power becomes reallocated in the process of overcoming the problem.

I would like to give a couple of illustrations. In Arkansas poultry litter is a major environmental issue for agriculture and water quality. The assumption behind federal policy, and how federal resources are allocated, is that it is a problem to be solved on the farm. Yet if we have to put all that poultry litter on these 40-, 50-, and 60-acre pastures in western Arkansas, we will have ground-water problems and surface-water problems. But federal policies are recommending that. As an NGO, we have the opportunity to help create market mechanisms in the private sector to distribute litter from areas where it's causing problems to areas where it can improve soil fertility.

We also have a situation where the large poultry companies would prefer not to see the growers organized for a variety of reasons. So, growers are isolated. The solution to their problem lies in other parts of Arkansas that have sodic soils that have very low quantities of organic matter and that need to be restored. Poultry litter is a resource if we could move it there. But to do that we have to organize growers and we have to create markets. We have to bring people together. There is a concern about doing that because of the reallocation of power that could result.

In contrast, in the Delta here in Arkansas, we have a group of low-income, primarily black farmers. And if you look at land allocation patterns in the Delta, they are on the most marginal soils, just as the poor are in developing countries. Their soils do not drain well. These farmers have successfully organized. In doing that, they have demanded from the Soil Conservation Service, the Farmer's Home Administration, and other USDA agencies that their unique problems in farming on these marginal soils be addressed. They have been successful at it. But there has been a reallocation of resources that has created political conflicts, and those are real conflicts.

As an NGO, we have the opportunity to help, not by saying that we need to reallocate power, not by saying to those poultry farmers over in the Delta, you must dispose of your litter in such and such and such a way, but by working with them to solve the problems. So, as groups arise, resources are reallocated to solve their problems and the formation of those groups results in the reallocation of power, which ultimately will result in a more equitable distribution of wealth.

I do not think we can deal with issues of sustainability unless we also deal with issues of equity. Sustainability involves intergenerational equity. Equity inevitably deals with allocation, distribution of resources, and benefits, and that means that we're dealing with political systems. As a result, sustainability is and always will be a political issue.

KENZO HEMMI
Toyo Eiwa Women's University

No one can deny that the political structure and uneven distribution of agricultural land ownership in many developing countries tend to degrade their agricultural resources. But I would like to mention an additional cause—the scarcity of statesmen in newly democratic nations. We can distinguish politicians from statesmen. Politicians engage in political struggle and adjust themselves to what people desire. Statesmen think about the future of the nation and educate society about the actions needed to ensure future prosperity, equity, and political security.

The spread of democracy into many developing countries in recent years has made political leaders increasingly preoccupied with meeting the varied, and sometimes inconsistent, demands of their people. People, mainly those living in cities, have easy access to information about political and economic life in other parts of the world through modern communications. Based on this newly acquired knowledge, people are increasingly demanding more freedom in their political life and a higher level of living. So politicians become preoccupied with day-to-day activities, and they don't have sufficient time to think about the future of their country. Moreover, they have almost no time to visit rural areas. When I travel in developing countries, I am often surprised that the government officials and politicians so rarely visit the countryside, so that they are unaware of resource or environmental degradation. They don't know what is going on in farms and villages.

If a nation has statesmen, it can make a great difference. For example in Thailand, the government, until 2 or 3 years ago, had allowed forests to be destroyed to increase export revenues. But the King of Thailand, looking to the future of his people, reversed this trend by starting a major Royal reforestation project. The reason he could do it, in my opinion, is he didn't have to reckon with politicians. He is somewhat insulated from the day-to-day politics.

Is there is a way to change politicians into the statesmen—to get them to recognize the resource degradation that is occurring in rural areas? I believe if developed nations, through international communication channels, make the political leaders in these countries aware of the fact of resource degradation and support them morally, financially, and technically, they will endeavor to protect their agricultural resources from degradation. They are extremely sensitive to prevailing opinion in the developed world, and their development budgets depend very much on financial flows from the developed world.

NEVA R. GOODWIN
Tufts University

The theme of the Rob Paarlberg's paper that stuck out to me was farm resource abuse driven by power abuse and, as an economist, that was a phrase that resonated because resources are, of course, one of the things that economics is supposed to be about and resources aren't supposed to be abused. The whole point of economics is that you have a market that is able to allocate resources to their optimum use, and that's obviously the opposite of abuse.

When you have farm resource abuse, you know there has been a serious market failure, and it shakes your trust of the market. People like Milton Friedman talk about how much they distrust government. They turn to the market as the alternative. And I think trust is the appropriate term to think about.

You have people arguing over and over again about which to trust more or distrust less, the market or the government. It's possible that there's a third sector that can help solve some of the problems that we've been looking to both markets and governments to solve. But if the third sector, the nonprofit, nongovernmental sector, can play such a role, we're going to have to rethink institutionally how it will do that because it's never been done before in an industrialized economy. Certainly, pre-industrial economies and societies had all sorts of nongovernment, nonprofit activities that worked well, but we're in a new world and we need some new institutional forms.

I've been thinking a lot about what I'm going to call the ladder model of development, which is intended to point toward a new institutional form that can take advantage of the third sector. We've heard a great deal about NGOs and the third sector. I believe the reason is, first, we have glimmerings of some new possibilities and, second, we've been bouncing back and forth between the government/market alternatives and feel dissatisfied with both.

We think of wealth, as in Smith's *The Wealth of Nations*, as resources to which human beings have laid claim, either by using them or by possessing them. The origin for nearly everything that we call wealth is the combined contribution of human labor with the gifts of nature. That's where you start. But to the extent that wealth is accumulated and people have more than they need for their immediate survival, you then have a surplus, and that's when life gets complicated and interesting.

William McNeill has described the group of people who figured out a particular way to do something about that surplus as "parasites." He lumps together protection rackets, tax collectors, emperors and builders of temples and pyramids, arbitrageurs, and merger-and-acquisitions specialists—all the people who see that there are small pools of surplus created by the people at the bottom of the wealth-creation pyramid and who pyramid it up into big pools of surplus that somebody else has their hands on. That process is what creates the political power that makes possible the sort of abuse that Rob Paarlberg was talking about. It also creates an opportunity. It creates the question: You have a surplus, you have more than you need, but what are you going to do with it? You're going to have more pyramids, more shoes, as Imelda Marcos did, more education. Or another option is what has sometimes been called philanthropy, or charity, or the third sector.

The last option—the third sector—assumes that once you've gotten wealth to pyramid up to concentrations, some people are going to have an interest in reversing that and sending cascading back down the pyramid to the people at the bottom who need it.

During the past 40 years, we've been in what we might call the first era of development assistance. I think it can be looked at as an era in which the idea was that there were rich countries and there were collections of surplus resources and that the rich should do something with these and send them to the poor. It has been an era in which there have been many fashions of thought. People have tried something and it didn't work, so they tried something else and it didn't work.

One way to look at the fashions of thought is that they've bounced back and forth between being top down or bottom up. You've had the sort of big dam projects, the big investment approach. You've had the basic needs approach. You've had the policy reform approach from the top. You've had human capital development from the bottom. None of these alone has been so satisfactory that people have been willing to go on indefinitely with them. They say, no, we have to try something different.

I believe we're now moving into what I would like to call the second era of development assistance in which the big donors—World Bank, USAID, the other multilaterals, and bilaterals—recognize that one of the basic problems with the first era was that because they were big donors with big amounts of money, they felt they had to move it in big chunks and, therefore, they had to move it to big recipients. It had to go in single chunks from the donor to a government agency or to whatever large recipient could handle funds in the millions of dollars. And not everybody knows how to do that.

Now it is being recognized that when that money spreads out down the pyramid, it finally has to get to the bottom where you have families, individuals, microenterprises, or villages, whatever actors are referred to as the grassroots. What is shaping the new interest in NGOs is that, in order to get from the top to the bottom, you must go through several different stages. You can't do it in a single jump. You must fill in what I regard as the rungs on the ladder between the top and the bottom. If you're going to rely on NGOs, you don't have to rely on them for the entire ladder, but they could fill in quite a few of the rungs. You could have a large global NGO that divides things up and gives part of what came from the original donor to various national NGOs, and it can go from them to smaller ones all the way down to the local level.

Let me give you one example. There was a USAID project on women and AIDS. It was decided to give the money to the International Center for Research on Women. ICRW had a competition, and they funded 15 or 20 groups around the world, which each got a part of the money that had originally come from USAID. One of the groups was in Senegal. I heard a lecture by a Senegalese anthropologist who described how they were further subdividing the funds they had received to two indigenous groups. The Dimba was a nonprofit group whose concern was fertility and early child care. The Lowme was a profit-making group whose concern was love and sex, and they gave advice to people who had problems. These two groups were well-established on the village level and had connections and trust relationships with the people who USAID wanted to reach. So, the money and the knowledge came all the way down the line. It got to these groups who were persuaded that it was worthwhile to them to learn something about AIDS because their clients wanted to hear it. So, that's a nice example of the ladder model working.

When we have this ladder model firmly in mind, we can conceptualize our own place. We ask ourselves: Which rung is my organization on? Who are the ones above and below me? How do I get to the people I want to get to? What rungs do I have to go through? I think that's a useful way of conceptualizing it.

Is this a way of reallocating power in the terms that Sandra Miller used? I hope so. If it is, will people put up with it? That's, of course, always the problem. And, is this one more Northern idea that we're going to try to impose in the South? When that question is asked, I respond by saying, whose South, which South? One of the most annoying things is to hear somebody say, don't tell us what to do about our women, or, don't tell us what to do about our poor or what to do about our minorities. The use of the possessive pronoun suggests that the speaker is not a woman or poor or a minority and has distanced himself from those groups, which makes me feel that the right of that speaker is no greater than the right of the outsider. And the outsider, if he or she comes from a group that has shared interests, may be able to meet on the same plane, coming in on a people-to-people basis or an NGO-to-NGO basis.

Rob Paarlberg rightly noted that the problems were initiated by European colonization. We could respond, then we had better not mess up anymore. Or we could respond that because our culture brought in the problem, our culture should take a hand in trying to help with a solution. I guess I would stand on the latter of those two.

Discussion

Harwood: I suggest that the basic problem is not so much a bipolar power struggle as it is a failure of institutions at both the local and national level to evolve at a rapid enough pace. Local communities often face rapid changes in their economic environment as well as population shifts, both up and down. Responses to those changes often require collective action, which, in turn, usually depends on appropriate institutional structure.

Traditional communities commonly have difficulty changing quickly. They often do not get clear or timely signals from central governments, which are often elitist and urban focused. Second, new problems may call for action from a new constituent group that may or may not have an effective institutional channel for either organizing or responding. Third, local systems evolve largely through experiential learning. In today's world, such first-hand experience is not always possible nor quick enough to allow a timely response to changes occurring outside the community. In some cases, such as with organic producers, self-imposed constraints on methods may preclude change.

I would suggest that rapidly evolving NGO organizations are key avenues for this short-term institutional response. They have the flexibility to respond quickly and effectively.

Paarlberg: It isn't the speed of change that we need to worry about so much as the direction. In villages in Africa today and in rural communities in Central America over the last 15 years, there was probably been more institutional and social change, and more change in relations within the family or between the countryside and the city, than anyone should be asked to accommodate.

Much of this change has come from population pressure or from the introduction of exotic technologies into what were essentially feudal systems. The solution isn't really to speed the pace of change. If anything, it has come too fast. Instead, the solution would be to empower those who are now being swept along or swept away by that change to give them a little more control over its direction. If you're a peasant farmer in a feudal system with only a traditional right of access to your land, and then a road is built and the land that you're on suddenly triples in value because it's good for producing cotton, you're gone. You're going to be pushed off. Someone is going to come in and rent, buy, or grab the land on which you have been producing the maize and beans that keep you and your family alive. Suddenly you're landless, and there was nothing you could do about it.

Harwood: But isn't the problem that you don't then have a local community institution that evolves quickly enough to handle that problem?

Paarlberg: In that sense, I agree. It's that institutional evolution that has to be rapidly accelerated.

Ellis: Dick Harwood said that people learn experientially and not really conceptually. Maybe that's true of organic farmers, but it's not always true of pastoralists. My experience with pastoralists is like my experience with my academic colleagues: Some of them are good conceptualists and some of them aren't worth a darn. But the pastoralists that I've worked with demonstrate a great deal of social and community-level adaptability—willingness to change their institutions rapidly when it's clearly within their interest. Yet they also can be extremely recalcitrant when they perceive that the change is not in their interest. But it's not because they lack the conceptualizing ability to change their institutions.

However, Dick is right when he says that experience is very important. I think that one reason that pastoral people are so adaptable is because they're highly educated in the sense that they travel around a lot and spent much time in different environments and meet different people and different communities. I think they can conceptualize what those different experiences mean for them, and they are able to adapt and change their institutions when necessary. But they certainly won't do it unless they perceive it to be in their interest.

Berg: This participation in power is a big one. In the *1993 Human Development Report*, the UNDP's view on market versus government in the Third World is, what's the difference? In many instances, both are in the hands of the elite, so the third way is empowerment. I would like to hold that up to more critical scrutiny.

I used to think that NGOs were more liberal than governments and, therefore, they would get more interested in empowerment. But, I think it could be the opposite. The NGOs may be a good deal more conservative. They may not necessarily be the revolutionary force that they're going to be asked to be. Second, I'm quite concerned that the Third World NGOs like to say they are in the development business because they do not want to say they are in the empowerment business, because they will blow their cover with their governments. I think we have to say that we're in the development business with them. The more we say we're in the empowerment business, the more we threaten their existence. Third, I think the United States is so heavy in this area that we ought to think of strategies that engage more of the donor community. We are half of the world's NGOs in terms of financial flows. Fourth, in the agricultural development area I think we're a little vulnerable to people saying, "Gee, how have you done in empowering your own small farmers in the United States?" or "Give us the lessons of that." Our closet is not exactly filled with shelves of answers.

Three questions: How do you foster large-scale change in the absence of a revolution? The East Asia cases Rob Paarlberg gave are right, but they're not necessarily the process that is going to take place elsewhere. We're going to see that, particularly in landholdings. Second, my organization is going to have a meeting on, "What are the billion dollar ideas in this field?" That is, if the Clinton administration and others really want to get into this, how do we take away from them the option of saying, "Let's do participation," and at the same time saying, "Great, now we can reduce the aid budget by 96 percent because we're only going to give $50 million to the African Development Foundation, the Inter-American Development Foundation, and a few tokens to NGOs"? And, third, how do you make this politically respectable, initially at least, to the governments of the Third World?

Paarlberg: On the first question, my reaction is to appeal to Neva Goodwin's very good argument that large-scale change isn't going to take place overnight in sweeping, revolutionary, country-wide or region-wide terms. Change will take place one village at a time, one river valley at a time, one household at a time, probably as a result of many separate and mostly private interactions. It will bear, in that regard, more resemblance to the workings of a marketplace than to a top down, centrally directed political activity or even to a bottom up, mass-based revolution.

On the third question of how to make this approach more acceptable to recipients and to clients, I agree you should not call it empowerment while you are doing it. Start by working to solve discrete problems that the governments in question are on record as having identified as needing a solution. Organize to hold these governments to account for what they claim to be concerned about, whether it's soil erosion or watershed degradation or salinization or loss of biodiversity. Search out the statements that they have made in various international forums or to their own NGOs, and then present what you're doing, or what you want them to do, in those terms, using their language. It may be that's all the cover that the more energetic, more creative, more innovative leaders of tomorrow within that government need to push their policy-reform agenda against the doubters of yesterday.

The second question I would have to think a lot more about. Other people might have a better answer.

Miller: The answer to your first question is embedded in your second and third questions. I think it's a mistake to assume that NGOs are the only answer. You can't invest large dollars in NGOs. The role of the NGO has to be clearly defined. There's a supply and demand for programs, just as there's supply and demand in the marketplace. Historically, we've focused on the supply of technology, the supply of research, the supply of capital, but it's all been supply. We've assumed that if the supply was there, the demand would be there.

What NGOs are uniquely qualified to do is to pull together the demand so that the supply is used more efficiently and to demand of the supply of programs that it be more responsive to particular needs. I think that when you're looking at the allocation of dollars, you must look at the mix of those dollars and where they're going, so that you have both the supply of information and technology and research and capital and irrigation and all these other pieces, but you have to figure out how you're going to make sure that the demand is there.

I lived in Botswana in an area where a large-scale irrigation project had been put in 20 years earlier. The water was used for municipal consumption and that was it. We had a beautiful lawn. There was virtually no industrial usage and there was no agricultural usage. I think that this gets at the issue that if there's no demand for those programs, we're wasting our dollars, and we must pay attention. That's why the role of NGOs is so important.

That doesn't say that they have to be grassroots NGOs. Some of that demand is grassroots, but there are problems that you cannot solve at the local level, where you need both national and internationally focused NGOs that are crosscutting issues. And, NGOs can work at a variety of different levels.

Goodwin: In relation to the question about how to make empowerment acceptable, one answer is, take the children of the elite and educate them in the West—as we've been doing off and on anyway—and make the idea of civic service a part of their education. One of the best exports that I believe that the North could make to the South is this idea of civic service—that if you have a big amount of surplus, you don't use it to build temples, you use it to make life better for other people.

Blake: I like your idea, but I would say we have a tendency to get a bit romantic about what we can do to help NGOs and to think of these things as revolutionary. After I started looking seriously at what really works, I found that a lot of my ideas were impractical. What works, works. That's silly to say, but I see so many different proposals that can't be financed. The only common denominator that I can find for things that work is firm involvement of farmers in projects that they believe are in their own interest.

There are three very basic questions that I would like to pose. One is, what can be done to help bridge the gap between governments and the people down below? When I think of bridging, I think principally of helping get governments off people's backs. The most useful things that an organization like USAID can do may be to convince governments to be more supportive of farmers, to make it possible for them to make available some of the services that farmers need, and to try not to run things at the local level too much. Another very tough question is, how do we get the right kind of technical advice to farmers plus whatever inputs they need? But the most important question is, how do we multiply the many successful small-scale farmer-conceived projects? How do you go from a small undertaking to enterprises that are economically and socially significant? How do you move to a larger scale?

Kaul: We should spend more time in understanding why some NGOs are more sustainable and resilient than others. Then perhaps we would do a better job. The Grameen Bank is a good example. Why has it succeeded when earlier attempts in Bangladesh failed several times? It must have some magic.

We should also not forget that a disproportionately large number of NGOs in the Third World are staffed by people who do not have a rural background. So, either you have urban people telling farmers what to do or people who live half way around the globe go around and tell people what to do. I don't think the solution lies there.

Tugwell: We're in a time of transition and that applies to our government as well as to Winrock. We have heard some proposals about what should be done to affect international development that are quite dramatically different from the earlier ideas. For decades political science has been worried about how to get empowerment without revolution. We have some suggestions. One is to educate the children of elites so that they will share power. I'm a little dubious about that. On the other hand, I think that we are perceiving a changed ethos in the world. I expect issues like participation and democracy may well extend beyond just the elite. So we can anticipate democratization processes spreading downward along with privatization, the sharing of political as well as economic power.

Also, it seems to me that we ought to pay more attention to investment of capital. Empowerment often means access to money. This is true in a gender sense, but it's also true in a social stratification sense. One way to get money to poor people is to make credit available at subsidized rates. In Environmental Enterprises, an institution that I've been working with, we've been doing precisely that. If you make sure that credit available to low-income people is at a rate that is reasonable and affordable to them, it may be one of the best ways to empower them and yet it is very indirect—indirect enough that the elites will not feel threatened.

Peterson: I'm not quite sure I agree that in order to solve the problems of sustainability and degradation you need to have empowerment. But I am concerned about how you might operationalize that. I don't really know how to go about doing empowerment. It seem to me that, in general, poverty leads away from power or power follows wealth, especially in nondemocratic systems. If that is so, then maybe I am involved in empowerment if I can find ways to increase the capacity of small farmers to produce more because that will make them wealthier and, being wealthier, they're going to, in some sort of natural way, get more empowered than they would be otherwise.

Goodwin: You get rid of the poverty and we'll worry about the empowerment.

Peterson: If that is the case, the argument isn't that I'm trying to do the wrong thing. The argument is, I'm not doing what I should be doing very well because somehow I'm not getting the wealth to get past the political side of it and get into the farmer side of it.

Smith: If you're trying to help the disempowered to move up, you must look at not only their educational level but how you can help them gain greater income-generating opportunities and achieve stronger decision-making roles. The three together make a powerful package that strengthens the status of the poor, particularly women.

Dietel: In a recent book, one of the gurus on the subject of corporate leadership says that the first responsibility of leadership is to define reality. What successful nonprofits do is to enable, assist, and facilitate the redefinition of reality on the parts of people who are largely powerless or have historically been powerless. And that is no small matter. When institutions like Winrock, foundations, whatever, go into these communities in poor areas—whether they're rural or urban—there is a transfer of those skills for managing the task of redefining the reality and reorganizing people in order to achieve what they regard as their legitimate ends. And, whether we do it consciously or not, in fact we contribute or fail to contribute in these societies to enable these people to do what they've already started to do but they can be helped to do it more effectively and, ultimately one would hope, to do it in partnership with the very forces of opposition within the society, whether it's government or business or both.

Paarlberg: I liked Bill Dietel's point that the enemy, to some extent, is fatalism and an inflexible view of possibilities. Political scientists know that people tend to organize for political action not to pursue what they want but rather to pursue what they think they have a chance of getting. It isn't the seed of "want" that needs to be planted. That is too easy. It is the seed of "possibility," the belief that effective organization and political action is possible. Then the problem of organizing for political action starts to take care of itself.

I liked Elise Smith's and Pat Peterson's comments as well. How to empower people while making it acceptable goes back to an earlier point that the education and employment of women is not only good for its own purposes and good for increasing control over resources and the sustainability of agricultural communities. It's also the best way to lower fertility rates and bring population growth under control. If you go into a country and say, "My purpose is to bring population growth under control," your efforts may pit this tribe against that tribe or this religious group against that religious group, and the political authorities may say, you can't do it. If you say instead that you're embracing their own goal, which is the education and employment of women, you will accomplish indirectly what it might have been difficult to accomplish directly. That's exactly the kind of strategizing that has to go on.

I also liked Frank Tugwell's comment that there's a chance that democracy will spread downward, and I think that has indeed happened in some Asian countries, including India, which was easily caricatured as an urban-biased political system 30 years ago. In part because of press freedom, rule of law, and considerable opportunities to organize even in rural areas, Indian democracy has gradually reduced the urban bias of Indian agricultural policies. Rural districts are increasingly powerful in the legislative assemblies. Some people would argue that in their price policies, at least, the Indian government is now rural-biased.

With the right political context, then, what begins as a democratic political process with advantages only for the urban middle class, can gradually evolve, by spreading downward and outward into rural areas. So, I think that is a point of hope.

The question that Bob Blake raised about multiplying change and about how that can be done forces me to back off a bit from the quick answer that I gave earlier dismissing the likelihood or the desirability of revolutionary change. In the political arena, revolutionary change is unlikely and, often, undesirable. But in the technical arena, such change may be absolutely essential. I've argued that significant political change may be a necessary condition for agricultural sustainability. I would quickly add that the other necessary condition may be rapid science-based technical innovations that allow us to produce more within a limited natural resource base. There is no other way, logically, to feed more people without destroying that resource base. Science-based technical innovation isn't something that can be done one person at a time by grassroots NGOs. It takes a commitment of public-sector resources to agricultural research, to national research systems, and to the extension of the fruits of those research systems.

We're talking about a long list of necessary conditions. No one of them is sufficient. But in the area of science-based technological innovation, the changes quite often are, in a precise sense, revolutionary.

Northrop: If we had the opportunities available, it would be interesting to think hard about USAID. This is an institution that could have a dramatic multiplier effect relatively quickly. It's about to go through a fairly significant change, we hope. Several of the things we would like to see happen could be catalyzed through USAID. We should all be pushing them to be thinking about building the future big vision for USAID

5
Forest Resources and Forestry Policies

Introduction

ROBERTO RAPERA
Winrock International

Today, we'll talk about forest resources and policy. The paper tackles several important issues and questions. These are:

- Forests are key elements in agricultural productivity through their role in watershed protection.
- Can forest resources influence rural productivity and sustainability?
- Can forest resources be assets in agricultural productivity and rural development programs, especially those targeting the rural poor?
- Can enhanced environmental qualities due to forest resources improve agricultural productivity?
- Donor communities will have to adapt their roles to the changing world—from working with central governments to working with NGOs and the private sector.
- Winrock's vision and strategic goals for its farm and village forestry program—for example, the focus on appropriate farming systems, multipurpose trees, and appropriateness of village forestry.
- Agricultural productivity and rural environmental quality are strongly affected by forestry.
- Forestry can be improved by policy and management changes. Crtical issues are tenure and property rights, taxation favoring forest conservation, market-oriented public and private timber sale mechanisms, and conflicts between timber values and biodiversity preservation and between forest values and environmental preservation.

I could not agree more with the authors especially in connection with their last observation regarding forestry and policy/management improvements. In the Philippines Natural Resources Program, in which Winrock is doing the policy monitoring and assessment component, the government uses policy reform to promote the management of forest resources. The policy areas addressed in this program are tenure security, timber pricing, forest industries, old-growth forest preservation, rationalization of residual forest management, technology development and transfer, and increased NGO and community participation in forest management.

The continued increase in global population and the alarming rate of environmental degradation are two major concerns that can be directly linked with forest resources and policies. Increases in population strain food-production capacity and lead to tillage of marginal lands that once were under forest cover. Population growth also requires that agricultural productivity per unit area increase continuously.

However, because of deforestation caused by clearing of forests in order to settle, house, and feed more people, environmental degradation sets in. Environmental degradation, as we all realize, lowers the quality of human life. It lowers productivity of all resources, including human resources. Whether forest resources and policies can impact on agricultural production and environmental quality is the topic of this session.

Impact of Forest Resources and Forest Policies on Agricultural Productivity and Environmental Quality

WILLIAM R. BENTLEY
Winrock International

CHARLES R. HATCH
Winrock International

JOHN C. GORDON
Yale University

Natural scientists have long considered forests to be critical elements protecting the agricultural productivity of watersheds. In recent years, farm and community forestry have gained prominence in rural development strategies because they provide direct subsistence as well as cash returns to rural people. Environmental organizations have brought renewed attention to global forest resources. Without question, forest resources are critical components of many farming systems worldwide and of the global ecosystem.

Given the importance of forest resources to both agriculture and the environment, can they play important roles in determining the productivity and sustainability of rural communities and economies? Can they be assets in agricultural and rural development strategies? Can they be useful in development strategies targeted to benefit the rural poor? Can enhancement of environmental quality with forest resources also improve agricultural productivity? In this paper, we provide some answers to these questions. We discuss the history of forestry and how we see the future. We consider the role of forestry in many current environmental debates, the roles of insiders and outsiders in resolving these issues, and the evolving roles of development donors in forestry. Winrock's mission identifies the rural poor as its primary beneficiary. As a consequences, we are especially concerned with some of the equity questions raised by the current environmental debates and movements. In the process, we expand upon Winrock's vision and strategic goals for its farm and community forestry program.

How We Got Where We Are

The history of forestry provides useful lessons about forest resource problems that developing countries face. In the past, many western nations experienced situations similar to those in the tropics today where existing forest resources are being rapidly harvested and deforestation is stimulating significant changes in national forest policies. Several factors are driving up the demand for wood, but especially populations, per-capita incomes, and urbanization. The consequence are real price increases for timber, solid wood products, and wood fiber, and many political arguments over allocation of public forestlands among various commodity and environmental ends. The political arguments, coupled with rapid increases in price, are compelling policy changes—public forestry services are moving beyond their colonial roots, and private property rights are changing to favor private forest investments. Except for rapid population growth, the same factors have influenced forestry policy in the West.

Forest resources: Major components of the global ecosystem

At the beginning of recorded history, high, or tall, closed-canopy forests, savannahs with trees and grass species, and scrub and dessert forest vegetation covered perhaps 60 percent of the earth's land area, the rest being open grasslands, deserts, tundra, and rock (Repetto and Gillis 1988). Today, forest vegetation covers about 7.2 billion hectares—two-fifths of the original 60 percent. The rest has been converted to croplands, pastures, and a variety of degraded landscapes that often produce little net biomass. Much of the forest area that remains has been cut over and some is seriously degraded. In some areas, the present forest or woodland has reverted from cropland or pasture. Closed forests, where little light reaches below the tree canopy, occupy about 2.8 billion hectares (Table 1) and contain 22.4 billion cubic meters of growing stock (Table 2).

The economic meaning of forest resource data is not always immediately obvious. Physical volumes and timberland areas are not equal in value in various global regions. For example, North America has almost 26 percent of the world's coniferous growing stock and 35 percent of the coniferous timberland,

TABLE 1
Land area (million ha.) of closed forests by world region, 1970s.

Region	Coniferous	Broadleaf	Total forest	%
North America	400	230	630	22.3
Central America	20	40	60	2.1
South America	10	550	560	19.8
Africa	2	188	190	6.7
Europe	75	50	140	5.0
Former USSR	553	175	765	27.1
Asia	65	335	400	14.2
Oceania	11	69	80	2.8
World	1,136	1,637	2,825	100.0

Source: Persson 1974.

much of it medium- to high-productivity sites. In contrast, the Newly Independent States (former USSR) have 60 percent of the coniferous growing stock (and 50 percent of the coniferous forestland), but the resources are mainly on low productivity sites in Siberia. Forest stands are isolated from markets and the volume per stem or log is low, which make roading and harvesting uneconomic. Since 1950, experts in the forest trade have waited for the deluge of Siberian timber. More volume may come from Siberia with movement toward a market economy, but the large timber inventory is worth far less than half as much as the volume in American forests.

In the developing world, Latin America and Africa each have 40 percent of the forest resources, and Asia holds less than 20 percent (Table 3). Over half the remaining closed tropical forest is in Latin America; Asia has 68 percent of

TABLE 2
Volume (100 million cubic meters) of timber growing stock by world region, 1970s.

Region	Coniferous	Broadleaf	Total forest	%
North America	265	230	360	16.1
Central America	7	15	22	1.0
South America	5	595	600	26.8
Africa	1	51	52	2.3
Europe	80	40	120	5.4
Former USSR	612	120	733	32.7
Asia	55	285	340	15.2
Oceania	3	10	13	0.5
World	1,028	1,201	2,240	100.0

Source: Persson 1974.

the tropical plantations; Africa holds 62 percent of the open or savanna forest/grassland types as well as over half the shrub and fallow forests.

Difficulties in interpreting deforestation estimates

Deforestation is both a loaded word and an ambiguous term. Instead of capturing the renewable nature of forests, deforestation connotes the ecological destruction of old-growth forests with the consequent loss of species and habitat. It suggests an image of greedy individuals and organizations pillaging our natural wealth.

Conventional measures of deforestation are limited to forest area harvested, and they do not either measure or predict what happens following harvest. In other words, deforestation encompasses the areas harvested as part of silvicultural cycles in natural mixed age and species forests, the conversion of old-growth to second-growth or plantation forests, conversion from forest to croplands or pastures, and the degradation of forests directly or indirectly into wastelands. All these circumstances are lumped together in current estimates of 0.9 to 1.0 percent annual deforestation (Gowen, Bentley, and Stijfhoorn 1994). This clearly is not a measure of what most people mean by *deforestation*. Currently, about half the area where timber is removed is immediately con-

TABLE 3
Estimated percentage of forest areas in developing world in the 1980s.

Region	Natural closed forest	Plantation	Open broadleaf forest	Shrubs and fallow forest	Total
Asia	26	68	8	15	18
Latin America	56	21	31	31	41
Africa	18	10	62	53	40
Total	100	100	100	100	100

Source: Derived from Gowen, Bentley, and Stijfhoorn 1994.

verted to cropland and pastures. Most of the timber is burned rather than harvested and utilized. Less than 15 percent of the area is harvested primarily for timber, about the same as the area harvested for fuel wood (Goodland 1990). Another 20 percent is lost to urban land use, roads and other infrastructure, or natural and human-caused disasters.

Using the FAO (1990) estimates from satellite imagery, Gowen, Bentley, and Stijfhoorn (1994) estimate that about 0.9 percent of the world's tropical forest area is harvested each year. The rate varies, however, from 1.2 percent in Asia to 0.8 percent in Africa (Table 4). While we do not have a precise estimate of the land areas actually deforested, we suspect over 50 percent of the area harvested returns to some form of forest vegetation.

One prominent way that agricultural development has affected forestry is that much of the current cropland and some pasture lands were formerly forested. This conversion has caused policy conflicts between foresters and agricultural development proponents. Some of these conflicts have intensified in recent years with the addition of environmental concerns to tradeoffs among food and timber commodities. In particular, the World Bank has been criticized for supporting projects that include forest clearing for agricultural purposes or for irrigation reservoirs (World Bank 1991).

TABLE 4
Estimated annual amount of tropical forest area harvested, 1980-90.

	%	million ha
Asia	1.2	3.6
Latin America	0.9	8.3
Africa	0.8	5.0
Total	0.9	16.7

Source: Gowen, Bentley, and Stijfhoorn 1994.

The United States has deforested a considerable area for agriculture. Clawson (1979) observed that about half of the United States originally was forested. He estimated that the area currently in the United States had over 950 million acres of forestland in colonial times, 850 million of which were "commercial" (defined as supporting 20 cubic feet of growth per acre per year). The commercial area dropped to a low of around 460 million acres between the world wars, but rose to 483 million acres by 1987 (Table 5). Noncommercial forestland occupied 258 million acres.

The conversion of forestland for agriculture has been reversed in most areas. In Connecticut, for example, 80 to 90 percent of the area had forest cover in colonial times. By 1840, when most lands were in pasture and fields, the forested area reached a low of 10 percent. Today, much of the cleared area has come back into forest. Trees cover about two-thirds of the state's area despite the enormous pressures of urbanization. A similar but tropical story is observed in Puerto Rico, which also went from nearly 100 percent forest cover in colonial times to a mere remnant of forestland, then back to about 70 percent today.

The U.S. timber resource, although considerably changed from colonial times, is growing rapidly and in most contexts growth exceeds harvest. Before World War I, for example, Chapman and Bryant (1913) warned that old-growth southern pine in south Arkansas would likely be cut out before second-growth timber was ready to be harvested. Yet 40 years later, in that region of the state, Georgia Pacific Corporation acquired the Crosett Timber Company, which had 1.3 billion board feet of timber despite continuous harvesting over the interim period. Georgia Pacific began a rapid harvesting schedule to repay the debt incurred for the acquisition and in the 40 years since then, it has harvested 2.6 billion board feet. Another 1.3 billion board feet

TABLE 5
Forestland area ownership and aggregate growing stock for the United States, 1987

Region	Total forestland	Area (million acres) Commercial timberland			Private commercial (%)	Volume of growing stock (billion ft^3)
		Total	Public	Private		
North	170	158	32	126	79.6	190
South	203	195	20	178	89.9	238
West	358	130	85	45	34.9	387
Total	731	483	137	347	71.8	756

Source: Statistical Abstract of the United States.

remain on the Arkansas property. Arkansas no longer has any old-growth pine to speak of, but the harvesting of southern pine has demonstrated the renewable nature of forest resources for timber production.

Ownership of forest resources

The United States is an interesting place to study forest ownership. About 28 percent of the commercial forests (and virtually all the noncommercial forests) are in public ownership, primarily national forests. Industry has another 12 percent, and 60 percent are in small, nonindustrial private forest (NIPF) ownership. For many years, the small private ownerships were viewed with concern by U.S. Forest Service leaders and other forestry professionals because the intensity of management was low, resulting in low yields per acre and low average value per unit volume. Industrial owners, on average, managed forests intensively and had the highest yields. Public ownerships were in between industrial owners and NIPF owners. More recently, however, NIPF management intensity has risen.

NIPF owners have probably been rational over the years in a direct economic sense. They faced relatively high interest rates, and given an aversion to risk, they cut earlier and invested less per acre than either industry or public agencies. As real prices have risen in recent years in response to scarcity of quality lumber, NIPFs have invested more, and they increasingly hold timber for longer rotations to benefit from quality and value growth.

In addition, more widely distributed ownership leads to more economic development from forest assets. Laarman and Sedjo (1991) refuted a 1960s theory of forests as development assets based upon the assumed successes of industrial forest management. That theory held that industrial forest development would create infrastructure and employment and would convert complex tropical ecosystems into development assets. In fact the results, with few exceptions, were ecological and social failures. Instead, Laarman and Sedjo found that economic development has progressed most rapidly in places like Scandinavia and the U.S. South where ownership is mixed and widely dispersed. The net benefits accrue broadly, and the cultural systems used are unlikely to cause degradation. Although less well documented, agroforestry or farm forestry systems seem to produce similar results in many tropical nations.

A hypothesis similar to the one that wide dispersion of forest ownership favors development holds for village control over commonlands and public forests. Social or participatory resource management systems that produce favorable results involve wide participation among villagers and collaborations among village associations, NGOs, and government forest departments (Wells and Brandon 1992, Lai and Khan 1993).

Forest policies: From different traditions than agriculture

Just as deforestation and land ownership patterns have evolved over time, forest policies have historic roots that are worth exploring. Forestry evolved separately from agriculture in Central Europe from Renaissance times onward, but retained many feudal traditions. In particular, the ownership of most forest resources continued to reside with the sovereign while croplands and pasture lands passed to the aristocracy and eventually to the peasantry. The U.S. forestry tradition evolved from German practices by way of British India. After the Mutiny of 1857 in India, the British began railway construction, and management of forest resources for sustainable sources of fuel wood and railway ties became strategically important. Because the British had no forestry tradition, they brought in German professionals to establish the Indian Forest Service. Dietrich Brandis, a German forester, served as the first inspector general of forestry for India, then was mentor and close advisor of Gifford Pinchot, the father of the U.S. Forest Service. German foresters, such as Shenck and Fernow, who had experience with the forestry college and research institute at Dehradun, India, moved to the Biltmore School in North Carolina, Cornell University, and other early centers of professional forestry in the United States. The connections between the American traditions, India, and other tropical nations is just beginning to be appreciated by forest historians.[1]

Two consequences of forestry evolving separately from agriculture are the strong inclusion of forest ecology into the scientific training and the integration across applied disciplines in professional curricula (Gordon forthcoming). Gordon and Bentley (1990) compared forest science and agricultural science to illustrate some of the differences between the two.

Forestry characteristics	*Agriculture characteristics*
perennial woody plants	annual or biennial plants
focus on vegetative growth	focus on reproductive growth
thousands of species	tens or hundreds of species
mixed culture frequent	monoculture frequent
multiple products and values	single or few production goals
extensive, low-cost cultivation	intensive, high-cost cultivation
positive wildlife values	negative wildlife values
negative domestic animal values	positive domestic animal values

The biggest contrast between forestry and crop production and most livestock production is that nature still is the primary competition for timber produced from managed forests. It is difficult to justify investments for 10 to 50 years or more to produce a product that other forests, often publicly owned, are selling for stumpage fees (prices or royalties charged for public timber on the stump) that barely cover the cost of sale administration, much less roads and other infrastructure. Rising real prices for quality timber, however, are bringing stumpage fees closer to the marginal costs of production.

In addition to general forest policy, research policy is also critical to the usefulness of forestry in development. Although a Green Revolution is occurring in forestry and agroforestry, sudden and dramatic changes in yield of desired products are not nearly as spectacular as those achieved in wheat and rice. One reason is simple leverage. Reallocating photosynthate from grass stem to grain head through breeding has much higher impacts than reallocat-

---

[1] This point has been recognized by a few Indian and American foresters for some time. The relationship is currently being studied by Edward Brannon, who is director of Grey Towers Historical Monument, USDA Forest Service.

ing photosynthate in branches and cones to central stem. Another reason is that commercial forestry has many more important species and genera than crop agriculture, and they must be collected, selected, and bred. Despite these disadvantages, a number of gains are possible from genetics combined with more intensive silviculture. Examples of success stories include Winrock's multipurpose tree species network, the New Zealand experience with Radiata pine (Sutton 1994), and the American high-yield Douglas fir and southern pine systems where more of desired products have been achieved through a mix of breeding and silviculture.

Policy and management improvements will be the source of many improvements in forestry. It is not happenstance that the new Center for International Forestry, being established in Bogor, Indonesia, by the CGIAR, will focus on forest policy. The International Centre for Research on Agroforestry devoted much of its first 10 years to diagnosis of farmer problems and designing solutions because so many agroforestry problems have social factors at their core. Improved tenure and property rights, market-oriented public and private timber sale mechanisms, and taxation policies that favor forest conservation are among the many policy and institutional areas that will have high payoffs from effective applied research.

Forest trees: Importance in agricultural development

The shift in development focus to the rural poor and farming systems approaches has led researchers to a better understanding of the roles of trees for fuel wood, fodder, and minor forest products in subsistence systems (Gupta 1983). From recognition of the value of multipurpose trees in farming systems two decades ago, we now have several national and international networks, including the Winrock multipurpose tree species research network in Asia. More attention is being given to trees as cash crops. Winrock is developing information on the rising real value of quality timber and integrating this information into its understanding of farm and village forestry systems.

Farmers involved in farm forestry can capitalize on the rapid increase in real prices for quality timber if they fully understand what the demand is and what cultural practices produce the highest value timber. For resource-poor farm families on lands marginal for food crops, farm forestry is an attractive option because usually these sites were once forested and can be rehabilitated to grow quality trees for fruit, timber, and other purposes. The existence of enforceable tenure or property rights are critical to allow the poor to capitalize on opportunities for tree growing, and usually some sources of intermediate-term credit are necessary.

Village forestry, which is practiced by groups that have defined access rights to commonlands or public forests, requires more complex social-managerial models than are necessary for industrial or public forest ownerships. However, where a group has already been formed for other reasons, such as a dairy cooperative, or where cultural unity is already high, as among many tribal or indigenous forest people, village forestry offers critical opportunities in rural development strategies.

Harvesting and processing of timber provide a good base for secondary employment opportunities. If quality timber is produced, considerable value can be added in rural areas. Because timber is heavy, the initial manufacturing and much secondary processing must occur in close proximity to forests to reduce weight before transportation to wholesale and retail markets. Sawmilling, one of the earliest forms of industrialization, can be labor intensive. In the United States, harvesting and milling productivity have improved rapidly over the past two to three decades, however, because many of the tasks are subject to rapid automation with computer-assisted technologies.

How We See the Future

Our perspective on the future is highly conditioned by our understanding of the past, but also we are influenced by the rapid increase in forestry's importance both politically and economically. Forest resources are central to many environmental debates, but understanding of development issues is changing rapidly, as is understanding of what development agencies can and cannot effectively do. Winrock has some important roles to play as an outsider for local people who are concerned with forestry development, and this clientele has influenced the design of Winrock's farm and community forestry program.

Forestry: On center stage in environmental debates

Historically, forestry has played second fiddle to agriculture in the United States and in most developing nations. With rising worldwide emphasis on environmental values, however, the balance is changing. Forest resources and forested landscapes are considered the last remaining wildlands largely untrammeled by human activity. The mythology of forests also is present in this appeal, be it the simple majesty of many old-growth forests or the darker images of the Druids and our heritage of Central European fairy tales. Sustained yield, in any case, can no longer be defined simply by an even flow of timber harvests. A broader conception of goods and services has emerged in professional forestry where the healthy forest is one that can produce an ever-changing blend of values.

Forest resources produce many services that are not priced by the market, such as watershed protection, recreation and wilderness preservation, and preservation of biodiversity. The first U.S. national park, Yellowstone, was set aside in 1872. Similar preserves of natural systems are common throughout the world, including the tropics. Environmental politics have focused public attention on reserving more rainforests and other vestiges of old-growth forest systems. Just as rising prices reflect economic scarcity, rising political signals and responding public support reflect the dwindling opportunities to protect many forest habitats and species that are endangered. This is a direct reflection of their rising scarcity and value.

Debates about forest resources are not confined to the non-market-priced services that forest resources produce, but include conflicts between timber values and preservation of biodiversity. Strategies to strengthen and develop national economies using products from the harvest of old-growth forests increasingly conflict with socially expressed desires to maintain the biodiversity of fauna and flora in these ecosystems. Nowhere is this conflict currently more apparent than in North America. But the conflict is not limited to rich countries. Recently concern that old-growth timber might be used to develop the economies of certain Newly Independent States immediately initiated internal debates among environmentalists about the values of their forests. Strong environmental movements are also present in countries like India, Indonesia, and Brazil.

Increasing attention is being focused on conflicts between agricultural development and the preservation of biodiversity. Initially, this conflict centered on increasing population growth and a nation's need to provide food security for its citizens in relation to the soil degradation consequences of the movement of agriculture from productive croplands to marginal lands and forested watersheds. Now this concern has expanded to include biodiversity issues ranging from loss of forest cover and the associated flora and fauna to introduction of farming systems that use pesticides or other chemicals that seriously alter predator-prey relationships.

Most recently, the impact of deforestation on global warming and carbon sequestration has attained international importance. Preservation of old-growth forests and tropical ecosystems and the implementation of massive reforestation campaigns are examples of actions flowing from global concern with deforestation. Tree planting is done through large-scale government programs and many small nongovernmental programs, such as Global ReLeaf organized by the American Forestry Association, which attract private and industrial funds and other kinds of support.

Conservation is, at its core, an issue of inter-generational equity. Parents who are concerned with conservation want to present their children with a better world than they inherited from the previous generation. Symptoms of rising public interest in conservation include the increased involvement and financial participation of individuals in conservation organizations.

Environmental preservation versus forest and tree management for goods and services is an argument about intra-generational equity. As such, it often is a rich versus poor issue. It is further complicated by the fact that in the past many nations environmentally mortgaged their natural resource base to attain industrialization. As the world has become a more closed community, this kind of strategy is no longer acceptable to many donor nations and recipients alike. Relationships among rich and poor nations already were strained. More rigorous environmental goals, but especially the preservation of scarce habitats and endangered species, aggravate the value differences that underlie conflicts between donor nations and recipient nations.

Understanding donor community roles in a new era

A pragmatic preference among both developed and developing nations for market mechanisms has replaced central planning as the favorite development paradigm. As countries industrialize, people increasingly move from rural areas to the cities. Consequently, political bases shift. These bases change yet more rapidly as market mechanisms and private firms are used to deliver agricultural services in place of public agencies. Often these new political constituencies attempt to broaden tax bases to include farm and forestlands. Collectively, the new political balance leads to more changes in land-use patterns than were caused by land reform initiatives in the past, but of particular interest are the forces leading to more widely held land ownership. One likely consequence of broader ownership is more rapid economic development. With inclusion of agroforestry systems into the working rural landscape, benefits accrue more broadly than before and the new activities are far less likely to cause degradation of the environment than the crop and livestock practices they replace. Certainly this is the result that has flowed from nonindustrial private forestland ownership in North America.

The ability of the donor community to require certain changes in rules or actions in recipient nations is more limited than it was 20 years ago, but it is especially limited in the current development era. Today's development paradigm focuses on markets and democracy. Donor agencies have to restructure themselves and their modes of operation to work effectively with NGOs, private industry, and private banks. Currently, the donor community attempts to target these groups through centralized government organizations and policies. The need for the future is decentralized programs that often operate almost entirely outside government. Whether the development assistance organizations can make this transition is not certain.

Funds and other donor resources are becoming more limited as a result of economic slow-downs in major donor nations. This decline is exacerbated by the need to provide development assistance to buttress democracy in the Newly Independent States. Provision of development aid to Asia is near an end, and development funds for Africa and Latin America are jeopardized by other priorities. In addition, past ideological and foreign policy justifications for development assistance are becoming less compelling as nations struggle to develop strategies in a new world order that is emerging from the cold war period. Furthermore, burgeoning democracy may increase the desire of recipient countries to be self-sufficient as broader participation in decisions leads to a stronger sense of national pride. One aspect of self-sufficiency is a nation's urge to have greater influence on conditions established with aid, international trade, and related country-to-country linkages.

In this changing and more independent environment, a donor's most effective role will likely be catalytic. That is, it will help institutions and organizations achieve their strategic objectives by providing dialogue, studies, and innovative pilot projects. Support of networks to disseminate new ideas and results and to develop human resources will increase in importance. Networks will be critical to meeting the continuing need for training in leadership, management, and technical subjects.

Insiders and outsiders have different roles to play

Most of the problems faced by a country must be solved by its citizenry if the solutions are to be stable and lasting. Insiders are best able to sort out complex relationships among social, political, and cultural factors that influence sustainable development. That does not imply, however, that outsiders' roles in development are unimportant.

The scarcity of trained foresters is a serious constraint in many developing countries. It means that few people understand forestry problems in causal terms and even fewer are able to search for solutions from local and international information sources. The human resource constraint is compounded by weak forestry, agricultural, and conservation institutions. This bleak picture prevails in Africa and is not uncommon in Latin America, and even some

Asian countries. Outsiders who are sensitive to local ecological and cultural conditions can help develop the human resources needed to create national and local organizations and to solve local problems.

Outsiders frequently can do things that, for political or cultural reasons, are difficult for nationals. For example, it often is easier for outsiders to look at new ways of overcoming old problems. Outsiders approach the problem with different biases and perspectives because they bring other experiences with them. This may also allow them to learn from other different but similar problems and how they were surmounted in other places. Outsiders often bring distance simply because they do not have a vested interest in the outcome; they are more independent in looking for solutions.

Outsiders are often able to build a broader coalition of interest to focus on a problem than might happen in their absence. There are numerous instances where wildlife, fisheries, and water quality issues in the Pacific Northwest of North America were resolved only after broad coalitions were put into place by individuals with no vested interest in the solution. Similar examples exist in Asia, Africa, and Latin America. Outsiders can facilitate compromise because they are not controlled by factions associated with the local political systems. To play the role of honest broker, it is essential that they remain independent of local interests and maintain their objectivity.

Outsiders may identify and help arrange financing for larger projects where it makes sense for all parties involved. This financing will often come from private sources and thus will not leave government or local private organizations strapped for cash to repay loans at a future time. For example, New Zealand farmers and European investors have developed "sharecropping" arrangements where the gross returns are shared at the time of final harvest of a timber plantation.

Outsiders can shield insiders from political and social risk associated with the introduction of new policies, restructuring of organizations, and similar changes. Insiders in the bureaucracies of many developing nations rarely expect the benefits to themselves from taking action to outweigh their expectations of penalties from taking the wrong action or even not taking any action. If insiders can position outsiders to be targets for fixing blame should the need arise, there may be significantly greater interest on their part in implementing change. Winrock's role is not simply to catalyze actions around new ideas, but to participate in the evolution of these ideas and, upon occasion, to serve as a lightening rod for the adverse consequences of development experiments that run amok.

Winrock's mission, vision, and strategic roles in forestry

Winrock's mission led us to a vision of its forestry program centering on farm and community forestry for several reasons. Forest investments can be made by resource-poor families on marginal lands. Policy and management interventions can skew the net benefits of farm and community forestry toward the rural poor.

Winrock's farm and community forestry program fits into its vision of itself as an agricultural and rural development organization (Winrock 1993). This agriculture-based program, however, has obvious relationships to traditional production forestry for timber and other commodity values and the newer environmental forestry, which provides watershed protection, biological diversity, and other services. These ties are essential if Winrock is to participate effectively in debates associated with global warming, carbon sequestering, biodiversity, and utilization of renewable resources.

Winrock has considerable experience in different types of development work with forestry elements. This experience provides an in-depth look at development issues in a variety of local environments, as well as holistic insights into development issues across regional boundaries. Winrock also has experience working with industry and farmer linkages to address development issues. Examples of Winrock's experience include:

- Several Asian pilot projects implementing agroforestry technologies and social forestry management systems.
- The Pakistan Forestry Planning and Development Project—a development project that has turned into a national farm forestry project that demonstrates how much farmers and industries can do working together and how much forestry agencies can change from being resource control organizations to farmer and industry service organizations.
- The multipurpose tree species network, which is helping Asian researchers focus on farm and community client needs.
- Facilitation of the forestry-environmental dialogue in Arkansas, which helped local groups form a consensus on clearcutting.
- Facilitating dialogues and conducting studies using forest economics and other applied social sciences to help policy makers and natural resource manager make better choices

Winrock is using its experience to design more effective uses of forest resources in rural development. These activities capitalize on the varied experiences it has acquired and derives strength from them. One dimension, tied to market mechanisms as a development tool, is directed at secondary wood processing in Arkansas and at wood-processing industry-farmer relationships in Asia. A second dimension builds on Winrock's roots in the Green Revolution and its more recent networking of science institutions. Both provide a springboard to address tree seed production and tree improvement worldwide, which is a technical need that will provide a long-term foundation for farm and community forestry. A third dimension focuses on studies, dissemination, and dialogue goals of Winrock's farm and community forestry strategy. It addresses farm forestry case studies from an economic perspective and timber price projections as they affect sustainable farm and community forestry development activities.

New Roles for Forest Resources and Forest Policies

This is a period of rapid change globally, and these changes are stimulating new understanding of forest resources and the products and services produced by forests. Old boundaries between agriculture, forestry, and the environment, both in the popular mind and among professionals, are breaking down in response to policy and managerial responses to new problems. In these changes and responses, we see a few points of particular importance to Winrock and the development community.

Recent environmental debates have brought more attention to forest resources, but usually have overlooked interactions between deforestation and poverty. Forest resources are critical to the global ecosystem, and they play important roles in determining the productivity and sustainability of rural resources. While environmental organizations worldwide have encouraged their supporters to be more concerned about forests, generally the stories presented overlook the costs of meeting preservation goals and in particular the costs paid by the rural and urban poor. Management of forest resources can channel benefits toward the rural poor, and increased production of timber products reduces the cost of both raw materials and fuel wood for rural and urban poor alike.

Discussions about development assistance are giving more attention to the different roles that insiders and outsiders play in resolving controversies surrounding forest resources and forestry's role in agricultural development and environmental quality. Donors and development organizations like Winrock can play catalytic roles in this new development era, including raising the consciousness of all players about equity questions.

Winrock is developing its mission into a clearer vision and more explicit strategic goals targeted on farm and community forestry. The vision includes a stronger role for private initiatives in forestry than generally are found in developing countries. Private initiatives include the many technical, managerial, and advocacy roles of rural-based NGOs. The new Winrock forestry program will help local groups play these roles more effectively. Farm forestry linkages to industry also are an effective means to sustain rural development in areas with wood-processing firms. Nonindustrial forestry at the farm and community levels is a mechanism that can increase economic development from forest assets and agroforestry systems, spread these benefits widely, decrease environmental degradation, and increase the area rehabilitated into useful land-use systems. The challenge for Winrock and similar development organizations is move this potential to a reality in several rural locations.

Literature Cited

Chapman, H. H., and R. C. Bryant. 1913. *Prolonging the cut of southern pine.* Bulletin 2. New Haven: Yale Forest School.

Clawson, M. 1979. Forests in the long sweep of American history. *Science* 204:1168-74.

FAO. 1990. An interim report of forest resources assessment 1990 project. Committee on Forestry, Tenth Session. COFO-90(a). Rome: FAO. Duplicated.

Goodland, Robert, ed. 1990. *Race to save the tropics: Ecology and economics for a sustainable future.* Washington, D.C.: Island Press.

Gordon, J. C. Forthcoming. Forestry: An overview with emphasis on the period 1950-1993. *The literature of forestry and agroforestry.* Ithaca: Cornell University Press.

Gordon, J. C., and W. R. Bentley. 1990. *Handbook on the management of agroforestry research.* New Delhi: Oxford and IBH.

Gowen, M. M., W. R. Bentley, and Erik Stijfhoorn. 1994. Tropical forest management and wood-based energy as development assets. In *Forest resources and wood-based biomass energy as rural development assets,* ed W. R. Bentley and M. M. Gowen. New Delhi: Oxford and IBH. In press.

Gupta, Tirath. 1983. *Economics of minor forest products in India.* New Delhi: Oxford and IBH.

Laarman, Jan C., and Roger A. Sedjo. 1991. *Global forests: Issues for six billion people.* New York: McGraw-Hill.

Lai, C., and A. Kahn. 1993. Participatory forestry in Bangladesh: First steps toward sustainability. In *Agroforestry in South Asia: Problems and applied research perspectives,* ed. W. R. Bentley, P. K. Khosla, and K. Seckler. New Delhi: Oxford and IBH.

Persson, Reidan. 1974. World forest resources. *Rapporter och Uppsatser* (Royal College of Forestry, Stockholm) No. 17.

Repetto, Robert, and Malcolm Gillis. 1988. *The forest for the trees? Government policies and the misuse of forest resources.* Washington, D.C.: World Resources Institute.

Sutton, W. G. R. 1994. The quality and economic implications of New Zealand's shift from an indigenous forest to a plantation wood supply. In *Forests and wood-based biomass energy as assets for rural development,* ed. W. R. Bentley and M. M. Gowen. New Delhi: Oxford and IBH. In press.

Wells, M., and K. Brandon. 1992. *People and parks: Linking protected area management with local communities.* Washington, D.C.: World Bank.

Winrock International. 1993. Farm and community forestry program. Photocopy.
World Bank. 1991. *The forest sector*. Washington, D.C.

Panel

BYRON T. EDWARDS
Board of Directors, Winrock International

I would like to narrow my discussion to what's happening here in the United States because I believe that some of the things that Winrock is interested in are coming in conjunction with what will happen to the forest industry here in the next 10 or 20 years.

National forests all over the United States are essentially being closed to logging. While many people out West are being put out of work and there are serious disruptions, the people of the United States essentially want their national forests and Bureau of Land Management lands preserved. So in the last 6 to 12 months, you have seen an enormous rise in the price of wood products. That reflects taking out of the system a huge volume of softwood timber. Prices in the United States are now at world levels. In Europe you seldom see houses built of the balloon construction techniques that are so common in the United States because the price of wood is so high that few can afford it. That will happen here.

Clearly, the principal benefit will come to the small woodlot owner whose timber has always been thought to be too scattered and too difficult to get out, in contrast to large woodlot owners—those with a 10,000-acre block, where it's easy to put roads in, and easy to manage. But with the price of wood rising you're going to see the small woodlot owner, the small farmer, the rural poor, gaining considerably more leverage in what they do with their timber.

You are going to see, over the next 10 years, the price of housing in the United States rise to 15 to 20 percent. We have not yet seen the full effect of federal timber being taken off the market. That will take another 5 years to work through. As in Europe and other places, you will see a change in house construction from balloon to brick and, particularly, post and beam, which will use hardwoods, which can't be cut so thin. That is the old style of construction in this country. Hardwoods once again will become relatively more important compared with softwoods, which are easy to saw and easy to peel. Hardwoods tend to be prevalent on small woodlots as opposed to large plantations, which are planted to softwoods.

You're going to see, particularly, an increase in harvesting in small lots in the Northeast. In New Jersey where I live, I see a sharp increase in cutting. You now see small woodlots being cut down, whereas 20 years ago you saw no skidders within 50 miles of New York City. And that is going to cause a reaction. Wealthy environmentalists will set up systems to prevent the small woodlot from being cut as much as their owners would like to have them.

On the other hand, in the South the small owner tends, when faced with the full range of economic scenarios, not to have his land cut all at once. He manages his land as best he can, much as a large timber owner would, getting a steady stream of income and a harvest level of about 5 percent. In other words, he has perpetual yield. So, if that continues to be true, with these higher prices, we are going to see a steady stream of wood available without over-cutting.

Also, as price levels rise, exports from the U.S. West are going to be stopped. The Japanese and the Chinese, who are waiting for timber supplies to come on line from their plantations, have been taking a tremendous amount of wood from the United States. That movement of wood is going to stop. They will have to get it from Chile and there will be a lot of pressure on the Chilean plantations until the Thai, the Japanese, and the big Chinese plantations come on line. I don't see them maturing for another 10 to 15 years, particularly the Thai plantations, which are geared to the Japanese market.

So there will be some wood shortages and high wood prices benefiting the domestic producers in the United States, particularly the small ones.

A word about the large producers. Most of the plantations in the South produce pulp wood. The growth in paper usage in the United States is going to be sharply curtailed, I believe. As everyone becomes computer literate less business paper will be required. We're already seeing that. As illiteracy in the United States increases, there will be less reading. The use of magazines will level off so that the paper business in this country will be stable. So, there will be some additional wood for construction but not enough to close the gap.

In Russia, the Central Siberian forest is now very expensive to cut. There are many small stems per acre, which are difficult for mechanized logging to handle. My Japanese colleagues have told me about their attempts to manage large timber concessions in Siberia and the work ethic problems that exist there. I think it will be 20 years before we see heavy production from that area. In any event, this timber will probably go to domestic markets because Russia is severely short of housing.

So, from the point of view of Winrock and what it can do and what will happen in this country, the emphasis likely will be on the small landowner, which is where Winrock has been concentrating. The United States will have to become self-sufficient from wood on small ownerships, which tend to be in the Northeast, the North Central part of the country, and to some extent in the South. We will see new styles of construction and a reliance on many different forms of timber and manufactured products.

CAROL STONEY
Winrock International

Bill Bentley pointed out that some of the main issues in forestry and forest management are productivity, sustainability, and equity. One thing that he didn't mention is the issue of stability in forest ecosystems, both natural and managed. The reason that should be included is that it's one of the four parts of the agroecological systems research equation that was proposed by Gordon Conway. I think what Conway meant is both the stability that occurs in the balances within an ecosystem, as well as the stability of a managed agroecological system in terms of the flow of products and benefits to the people who use the ecosystem.

An undisturbed natural forest is stable, even though it's constantly changing. The balance of predator and prey, of nutrient flows, and of water flows are in a stable flux. In an intensively managed ecosystem, like some agroecological systems, some of the problems have been where the situation becomes unstable. For instance, in Indonesia heavy use of pesticides killed a species of spider, which was a main predator of the brown planthopper, a rice pest, and those populations got out of control. Then the hoppers became resistant to many of the pesticides themselves.

So stability is a critical issue in terms of managing ecosystems. Even plantation forestry is a complex ecosystem, and we need to understand these issues, both in the short-term seasonal sense in the flow of products and benefits, for instance from an agro-forestry system, but also in the long-term sustainability sense.

It's a key concept in terms of some of the conflicts that Bill talked about that have occurred over how the land is used. These conflicts concern whether the forest should be just a natural reserve, whether it should be harvested for forest products, or in some places whether it should be cleared for some other use. These conflicts are occurring in this country and in many countries where Winrock has been working.

The most compelling argument for conserving forests as natural, undisturbed ecosystems is the conservation of the biodiversity resource. In this country, that argument has been used by environmental groups that want to prevent people from using the forest at all. Often the users and the environmental groups don't seem to be able to reach a compromise or don't even seem to be interested in talking to each other.

I find it interesting that in Asia most of the environmental groups that I came into contact with considered protecting the rights of forest users to be as much of their mandate and mission as protecting the forest was. The two went together. They weren't separate issues as far as they were concerned. But they were really talking about forest-dwelling people communities that had traditionally lived in the forest and did not have much decision-making power over how the forest was going to be used once the government or the private corporations decided to use them.

I think that there's a role here for organizations like Winrock to be a broker between the forest users—whether they are the forest-dwelling peoples or the private sector or the government—and to help those people work out ways to conserve and use the forest.

I see that traditional forestry is going to change, and it is changing, into new ways of managing land. It will affect other kinds of land use and management, not just forestland use and management. Much of this will happen at the margins and interface between agricultural lands and forestlands. We're trying to come up with solutions that enable people to get the products and the benefits they need from the forest, while at the same time the core of the forest can be protected and maintained in its natural state. To do this, forest management has to be participatory. That means that the communities surrounding forests must be involved in the management plans and decision making.

But who should be involved? How do you involve people in these decision-making processes? We can't just go in and tell people that things are going to work the way that we see as best way for them to work. We must let people decide for themselves.

One of the reasons many us got involved in social forestry and agro-forestry in the first place was the idea in the 1980s that there was a major fuel wood crisis, especially affecting women, who were the ones who collected the fuel wood. When I lived in Africa, that was mostly true. Women were the ones who collected the fuel wood unless the fuel wood was being collected to make charcoal and sold, in which case it was the men. At a workshop in the Philippines in 1989, it was asked, is it men or women who collect the fuel wood? They couldn't agree because there were people from all over the Philippines, and it's not the same throughout the country. So, you can't draw conclusions about who collects the firewood or who will benefit from planting trees.

That's not what we're supposed to do. What we should be doing is helping those communities figure that out themselves: What they need, who should plant it, who should maintain it, who should harvest it and sell it, or burn it, or whatever. One thing we can do is point out to them why all members of the community should be involved in the discussion, the women and the men and the elders and the young people. But we have to know the answers. We have to be thinking about these issues and learning more about what the gender roles are in managing the resource.

JIM WIMBERLY
Winrock International

We've talked about agriculture. We've talked about environment. We've talked about rural development. But there's another dimension that cuts across all of this. That's energy. Energy is critically important. It's required for all aspects of our agricultural production and processing-all aspects, even, of environmental management.

Right now, this planet is facing rising demands for energy. Even per capita demand for energy is rising, particularly in developing countries. Yet, we're entering a period where our traditional supplies of energy are going to decline. We are depleting our fossil fuel resources. So we're going to have to look at what our energy options are.

The 21st century is going to be a big century for energy because of the increasing demand and the decreasing supply of conventional energy sources. So, what are the options? There's a lot of coal in the United States and in China. There's a lot of coal in other parts of the world, but that brings with it many problems. There's a wide range in the quality of coal, and we get into many issues, particularly air emissions and other environmental problems, with continued and expanded burning of coal. Let's look at natural gas and crude oil. Those resources are also limited and they are being depleted. There's much political uncertainty associated with the supplies. So, we must consider other options.

One option is nuclear power. A huge proportion of the budget of the Department of Energy has been focused on nuclear power for many years. But there are still many unresolved issues associated with nuclear power. In particular, the storage of the spent fuel is not yet resolved, and there's no promise that it will be in the next few years. So what are we going to do to fill the gap?

From many perspectives, certainly from an environmental perspective and from a sustainable agricultural systems perspective, renewable energy offers a lot. But let's break that down and look at the different components of renewable energy. One that has been promoted a lot is photovoltaics. This is a good source for specific applications, but is generally expensive and has limited applicability. Photovoltaics will not fill the gap for this world in the next 10 to 50 years. Another one might be hydroelectric power. But, we've been pursuing hydro aggressively for many years and there are not many suitable sites left, and new hydro systems are being challenged by environmentalists. We hear a lot about wind power, but it's even more limited than photovoltaics. There are limited sites, and the total amount of potential energy from wind power is small.

So what's another option? Biomass. Biomass energy is a huge potential resource. Biomass is, in the broadest sense, almost any plant or animal material. In the world of energy, wastes such as sawdust, animal manure, or rice husks have typically been the sources of our biomass energy. But there are other sources. We are just beginning to see the potential for energy crops—actually growing fields of energy. There are still some technical hitches that need to be addressed in the production, processing, and conversion of biomass into usable energy, but there is an enormous potential here.

We're at the point now where it's a question of economics. Part of the economics has to do with how much of the external costs are factored into conventional fuel sources so that we have a level economic playing field. We're already starting to see that happen. The BTU tax the new administration proposed was a step in that direction.

The Electric Power Research Institute, a nonprofit, autonomous organization that does research for utilities, has spent 2 years analyzing the potential for biomass energy in the United States. They have concluded that, if the utilities buy into this new energy source, the potential is 50,000 megawatts by the year 2010. To put that in perspective, the total electrical capacity in the State of Arkansas is about 3,500 megawatts.

There is interest in energy crops at USDA, DOE, and EPA—not just looking at residue streams but looking at energy crops. Most of the people that are looking at this consider short-rotation woody crops, such as hybrid poplars, or perennial grasses, such as switchgrass, as the most promising energy crops.

This new arena could have a major impact on agriculture worldwide, not only in the United States. Developing countries can benefit from biomass energy and from energy crops as much as the United States, if not more, particularly in terms of percentage of installed capacity. So, we as agriculturalists have to learn more about energy and the energy, people have to learn more about agriculture.

Who will take leadership in this new industry? Utilities have virtually no experience with agricultural systems. Besides, they have limited mandates because they're regulated by the public service commissions. Government agencies can help, but I don't think they can take the lead. They are helping to stimulate the new industry, but the private sector certainly has to be involved. But can we rely on private firms to finance all the necessary research and development? Can we rely on the private sector to develop a complete new industry with adequate consideration for all of the environmental, equity, and rural economic development issues?

That leaves nonprofits or NGOs. In this arena, I think that there's a great niche for a strong nonprofit like Winrock International. To the utilities and the public sector, Winrock has credibility and, yet, we don't have conflicts of interest associated with private firms. On the flip side, the private sector likes us because we have the technical credibility, but we're not regulators and we're not bureaucrats. So, we are positioned to take a leadership role in this emerging industry that could fundamentally change agriculture and forestry, and further promote sustainable agricultural systems.

Discussion

Crosson: I heard Tag Edwards say that the big run up in timber prices over the last year and a half has been a reaction to the actual and, more likely, the prospective closing of public forests to exploitation. He went on to say that a principal beneficiary of this would be the small, nonindustrial holders of forestland, meaning, I would take it, that production from that source would increase. That leaves a question: Is that the only supply response that can be expected to come from these steeply higher prices?

Edwards: There will be a variation in timber pricing, but the run-up in prices has taken place during a period with no increase in home construction. In this country, we are, demographically and because of the recession, at a fairly low level of single-family housing starts. If that should increase from a level of 1.1 million to 1.5 million, I can't imagine what the price of wood would do.

But where would other supplies come from besides the small woodlot owner? The major corporate woodlots or plantations are generally going to pulp and paper, particularly in the South. But there has been a modification in the size of plantations.

The forestry profession within the private sector has been affected by the environmental movement. We used to plant wall-to-wall pine as far as you could see-in watercourses, draws, and stuff like that. We would knock everything down and start over with pine, which created an ecological desert. To be blunt about it, nothing could live under a closed canopy of that type. Today, you have much different plantations, but you have less softwood. You have hardwood shelterbelts. You have corridors for wildlife. So, you may be getting less wood from that sector, and it's also going to pulp.

You're not going to get wood from imports. The Japanese will bid us up as long as they have a strong yen. Canada is in a similar situation to the United States. That used to be our big source of extra supply. The Canadians are now going through the same thing with their British Columbia forest that the United States went through with its western forest a decade ago. There's enormous objection to cutting the last remaining stands of that timber.

On the other side of the Rockies in Canada, where you have some very large forests, you've had considerable exploitation. In eastern Canada, they've hammered those forests so that they're in worse shape than anything we have. So, you're not going to get much wood from Canada. The U.S. West is basically government land, so you're not going to get it there. You're not going to get it from anywhere else. You have to get it from the smaller and medium-sized owner in the Northeast and in the South.

Bentley: Timber trees present an interesting picture for forest economists because prices have been rising in real terms for a long time. Reliable data on lumber suggests real price has increased since 1800. Stumpage, the word for standing timber, has gone up in real price since at least the American Civil War. At Resources for the Future, Potter and Christy published a book (*Trends in Natural Resource Commodities* [Baltimore: Johns Hopkins University Press, 1962]) in which they looked at the price changes of different resources. They had expected to find that the nonrenewables were going up in real price and that the renewables were falling or staying stable. The fact that timber was going up in real price and iron and aluminum were falling was an interesting empirical discovery. It, of course, allowed for a lot of discussion of technology and institutions and substitutions and all the different things economists like to talk about, but trees kept standing out as being odd.

In the 1980s, Peter Berck at the University of California and I did a study of the old-growth redwood resource. The Congress of the United States then had decided to take half of it out of production one fine day in 1978 to create the second take of the Redwood Park. It was an econometrician's dream. It was quite clear that not only did the short-term supply of timber to the market respond to these changes in the inventory, but that the price-expectation models worked. It was reassuring to those of us who made our life study economics, I guess.

About a year ago we started to do some work in Winrock on timber prices. Some of this has been financed by the Internal Revenue Service of the United States, which is having an argument with a taxpayer in Alaska. We've managed to run down price data on timber and logs worldwide for long periods, and one thing becomes fairly obvious: If you look at short periods in the timber or log business, you get all sorts of stuff—pipeline shortages, mill shortages, etc. So you must look at longer periods of time—20, 30, 40 years—to get stable trends that you can explain in terms of inventory and so forth.

It appears, for reasonable quality material—by quality we mean you can get a 2 x 4 from it or some similar solid wood product as opposed to simply fiber—prices are going up 1 to 3 percent year. One of the conclusions of the work so far, is that the demand is for quality. It isn't just for fiber. It isn't for fuel wood. We don't have good information on all the nontimber values, be it fodder, knots, medicinals and so forth. I'm just talking about central stems—material that makes good sawlogs.

It would also appear that at least part of this trend is evident because we started these data series about 1800 when there was an enormous, worldwide, natural bounty, and the price has to rise to the point where it makes sense to invest in growing this stuff. And that's what we're seeing.

My Marshallian economics of 30 years ago says, when we get there, prices will begin to stabilize, subject to increases in demand. In the meantime, our concern is how this fits into world development strategies. What we have in timber is an investment that's going up in real value if you can stimulate farmers and villagers to grow quality as opposed to junk. Don't think they're going to make money out of just growing sticks to throw onto the burner. There may be other good arguments for doing that—biomass energy and such—but you're not going to help poor people doing that.

Along with the real price increase, you're getting 2 to 3 percent, or more, annual biological growth. So, this is an investment returning 5 to 7 percent in real terms and, where you're dealing with a high quality product, even a little higher. That's a very attractive return.

The short-term price changes that we're looking at now are run-ups in spot markets, the public timber markets, and the open log markets. They are short-term and transitory phenomena. They're very real if you're trying to build a house right now, but they probably are exaggerating this whole set of scarcity issues.

Timber scarcity is something that we're going to continue to study at Winrock. If we don't understand scarcity, we're either going to underestimate or overestimate the returns that can be made by putting trees into the rural landscape.

Coward: I'm interested in what the impact might be for small holders—how the gains that might come to those individual landholders might be transformed into economic development at the community level. Are there things that might be undertaken that would lead to value-added kind of activities within those communities?

Edwards: The processing of forest products or timber has been a fairly expensive operation. I think that will change. At the panel level, plywood will probably be replaced to a greater and greater extent by wafer boards, and that does require expensive processing and a fair amount of local timber. It's unlikely, I would think, that communities would get involved in that, although those plants will become more prevalent in areas where they are not now. The wage scales at those plants are fairly good.

In sawmills, where the wage scales tend to be lower and the technology runs all over the place, I believe in the next few years we will see a miniaturization of the technology that's in big sawmills. This will enable a small sawmill, say 5 million board feet a year, to be run by five or six people and to process the wood fairly close to where it is grown because transportation costs on wood are fierce. Those systems are being developed in Canada now.

Coward: What are the possibilities with regard to minor forest products?

Bentley: The term developed in South Asia during the British Raj for all the things that weren't timber, from fodder and fuel wood to medicinals, nuts, oils, all sorts of things. In India over the last decade or two, so-called minor forest products, in aggregate, have been more valuable than the timber. There are a couple of reasons for that. One is they cut down a lot of their trees. They can't make as much money from timber as they used to. Another is that we are seeing a shift toward some traditional products that come from these forests. As population and incomes grow, these same products are becoming less available for the same reason that natural forests are going down in availability, so the prices are rising.

These products can be important to a rural development strategy. Like fruits and nuts and other kinds of things, they are less prone to being part of a global market. However there is some evidence in Latin America that global markets can be established for some of these products, and then they have to be fit into the same picture we now can develop for timber.

In the future, however, we are going to see two things. First, most of the values that are associated with extractive reserves, minor forest products, and such will have to be domesticated. Extracting stuff from the natural forest and keeping a forest standing is hard to do unless you have a management system, and as soon as you have a management system—be it village controlled, farmer controlled, corporate controlled, or whatever—you have incentives to domesticate the system. Second, as soon as people can grow trees, rather than go out to a commons and cut them down, I think there is going to be considerably more attention paid to growing of trees for value.

I don't want to ignore the many gender issues between subsistence and cash within the household, but it's interesting, in some of our work in India and other parts of South Asia, that the initial demand for indigenous fruit-bearing and other tree species shifts toward timber species as people develop their farm and social forestry systems. It shifts that way because that's where the markets are pointing them.

The key is not to be doctrinaire about timber or minor forest products but to try to understand how these markets are working, the role of these products in subsistence households, and what the balance point is that makes sense for improving poor people's welfare over time.

Stoney: In the Java Social Forestry Project, which was funded by the Ford Foundation, we tried to figure out what plants farmers could grow in the forest, under the canopy. Most were medicinals and perfume oils along with some food products, but they didn't turn out to be very economical. The thing that worked best was the plant that's used to make patchouli oil. The farmers were able to set up their own steam distillation process. At first, they were just selling the plants to a local factory, but then they started making the oils themselves.

Ruttan: In this discussion, what I hear is that forestry is beginning to make the kind of transition that livestock made when we began to move from self-supporting livestock on natural pastures to fed livestock. We didn't get any productivity growth in livestock until we made that transition. At the same time, I'm a little worried about the direction of Jim Wimberly's comments, partly because of this competition for resource and partly because it seems to be moving back upstream. The long-run trend in energy production has been from high carbon to high hydrogen content. It seems to me that you're saying we're going to move back up that stream. I worry that if we try to move up that stream in anything except some small niches, we're going to be increasing the cost of both energy and food production.

Wimberly: One of the most attractive aspects of biomass energy is a net zero carbon cycle. That is, biomass systems do not add carbon to the atmosphere. From a greenhouse gas or global warming perspective, this is particularly important.

Competition for land use will certainly be an issue, particularly for multipurpose tree species because they can be used for a variety of proposes as well as energy. One key issue will be: What kinds of lands would be most suitable for energy crops? The emerging biomass energy industry is focusing on good agricultural lands because yield of bioenergy crops will also be very important.

There's also many environmental benefits associated with biomass energy, relative to other energy options. While it is true that the net energy return—the amount of energy obtained from biomass relative to the amount of energy invested—in some cases is not as high as we would like it to be, but it's still there. For most bioenergy systems, it appears to be a net positive energy balance.

Goodwin: Photosynthesis uses the sun's energy, and plants also draw nutrients from the soil. If you return to the soil what was taken from it, you have a stable system. But if you're burning or taking off and using all the biomass, don't you have a declining soil resource?

Bentley: It depends on what you burn. In trees the cellulose material doesn't store much nutrients. As soon as you take the leaves and roots off-site, however, your actions lead to rapid site degradation. This is very different than growing, say, sugarcane or one of the other C-4 crops for energy, where you're hauling most of the plant off-site including the leaves where a lot of nutrients are stored. The key is whether you're hauling nutrients off-site or not.

Wimberly: That's true. Most of the systems that I'm aware of are looking at only hauling the log itself and leaving most of the leaves and stems in the field. That's why I think there's a role here for Winrock is to help make sure the sustainability issues are not overlooked.

Goodwin: Carol Stoney mentioned that in the mid-1980s there was the perception of a fuel wood crisis. Was that only a perception? Has it gone away?

Stoney: I don't think it's gone away and maybe it's even gotten worse. In West Africa most of the village and community woodlot projects were started in the late 1970s and early 1980s. The idea was that they were to be for fuel wood for the women, especially. But when I visited in the late 1980s, many of the small stands that had been planted were being harvested for timber, not for fuel wood. I think there still is a need for fuel wood, but you can get cash for timber. And that's what people wanted more, I guess. There was a perception that women's labor in collecting firewood was not worth much anyway.

Goodwin: So they've gone on doing it?

Stoney: Yes.

Bentley: Where people have let those trees get bigger, you get a lot of fuel wood as a by-product. In Gujarat and some other places, it would appear that social forestry and farm forestry produced more fuel wood than they thought, even though they had moved them onto timber-sized stands.

Havener: I want to ask Sandra Miller to describe Winrock's work in Arkansas to develop a secondary wood products industry.

Miller: We've identified 700 secondary wood products made to measure in the state—everything from pallets to furniture. There's a tremendous opportunity to add value to the products that those manufacturers are already making. There has been fairly steady growth throughout the 1970s and 1980s in manufactured wood products.

At the beginning of the century, Arkansas was a leader in that industry. It's losing market share to other states because we haven't invested in that industry as we should. One of the things that we're trying to do is pull through demand to get better management, particularly of the private, nonindustrial lands so that we can get prices up. By adding value with these manufacturers, we've put together a for-profit sales company that's bringing new sales to Arkansas and then identifying manufacturers working with them and in the process adding employment. Interestingly, the Delta is where we've had the greatest success in adding employment and generating sales because there are many first-generation manufacturers there, and they're looking aggressively for sales.

Northrop: In discussions like these, I often hear a commodity focus. I think there's a distinction between the kind of forest that is only a commodity and the kind of forest that holds an array of biodiversity values as well. In temperate areas, I think about that as being old-growth forest. In tropical areas I'm not sure how to describe it, but I would suggest the Andean and Amazonian tropical rainforests as examples.

I have a concern for our planet's ability to maintain some of these systems that hold biodiversity values. So often in discussions about forestry I don't hear the distinction that allows for the planning to occur that will allow us to set up management regimes for the protection of these biodiversity treasure troves.

I wonder whether economists could think a bit harder about separating out those two types of forests, and then consider what that means to the global demand and supply of wood products. By beginning to understand whether there are wood needs that arise as a result, given that you hive off a part of your global wood supply, we could begin to think more rationally about the need for demand-side management of wood resources and management of new second-growth supply, etc.

I would suspect that you would have to do that for several scenarios because I don't think that the environmentalist's dream of preserving all of the biodiversity-rich forestland, globally, is going to come true. But there are probably gradations of areas of that supply of wealthy land that you would like to preserve. So, if we save 50 percent of it, what would that mean to the world supply of harvestable wood and what would that mean in terms of world management and planning for commodity supply, and so on?

Bentley: From a supply and demand standpoint, I am entirely in favor of all the reserves you can create because getting naturally produced wood off the market, locked up in one form or another, will lead to higher prices more rapidly, which make rural development investments, big and small, more sensible. That is the single biggest reason that, as we look to the future of public forestry in the developed and developing world, I just don't see the volumes being there. For the remaining quasi-natural stands, be it the old-growth forests of North America or whatever, there's going to be high demand in the political sense of the word to put them into reserves. The rest of it will get mismanaged in most of the developing world because they can't protect it.

There are two issues that we need to look at carefully. One is that the price and other terms of real dearness of this scarcity is going to be paid mostly by the local people, the marginal farmers near natural forested landscapes or the indigenous people. And they are going to pay simply because it will hit them and their livelihoods the hardest. I think we need to be conscious of that. Groups like Winrock have got to start working even more actively with the conservation and environmental communities. We need to become more active because the folks on that side are becoming increasingly conscious that if they don't solve the rural poverty problem, they aren't going to be able to protect these reserves.

That said, we also need to take a hard look at the biodiversity issue, and encourage more research on it. I recommend the review in a recent *Science* magazine of Ed Wilson's latest book. Its point is that the jury is still out on biodiversity—on whether it makes a bit of difference whether we have 8,000 kinds of ants on a hectare of land. I think the reason we aren't having this argument in agriculture is we already solved biodiversity by doing away with it. I don't think that that says forested landscapes ought to be domesticated and have the same thing happen to them, but this is why we're focused on wildlands, forested lands, and other lands we haven't figured out how to overly domesticate yet.

Harwood: It is true that we "solved" the basic problem of food production at least for a time, with modern technologies. We now have a broad range of secondary problems to overcome. The lack of stability in the balance between crops and pests and the lack of field containment of plant nutrients and of crop and animal residues threaten that production base. A good part of the answer to those problems lies in management of entire landscapes and the spatial diversity of crops and animals within them. We have not yet figured out how to retain the production gains of the Green Revolution, while achieving ecological stability in both a biological and sociological sense.

The physiological assumptions of some of our modern technologies are also in question. We employ an energy--inefficient process to produce carbohydrate, a "high value" energy source, using corn as crop, then convert the carbohydrate to fuel for combustion. We spend more energy in production than we ultimately recover.

Agriculture of the future will be subjected to far greater demands than simply feeding people, important as that may be.

The demands for efficiency of production in a resource-limited and populous world and the requirements for ecological balance and for low environmental impact add significantly to agriculture's burden. As biotechnologies expand the potential for agricultural production of industrial feedstocks, the burden on land will increase. Jim Wimberly is absolutely correct in that regard.

Tugwell: There has been a quiet and far-reaching change in the utilization of biomass resources in the United States. Most people are not aware that today we get more energy from biomass than we do from hydroelectricity.

A related point has to do with the issue of conversion technology. What is in the offing is a dramatic increase in the efficiency of conversion of biomass to electricity through the use of advanced gas turbines. These are second and third generation turbo-generating systems that are hooked to gasifiers. Work is going forward now that looks like it will make it possible to take large volumes of biomass and convert them to electricity at rates that are very competitive.

For those industries that have large amounts of crop residues, like sugarcane bagasse or rice hulls or forest byproducts like sawdust or other waste, it looks like these markets will be joined.

It also is beginning to appear that cultivated biomass—grasses and grasslands, especially marginal grasslands—are a more important resource than even waste wood. One concern is the market impact. Does it create a market for large plantation operations? Or does it create a market for small woodlots in developing countries, and what would the impact be on the natural forest?

Peterson: Carol Stoney mentioned the stress between the agriculturalists and the environmentalists about protecting resources as opposed to managing them. Is there a role that agroforestry can play in this because they do deal with that issue of managing resources for their protection, as opposed to sort of putting a fence around it and keeping exploiters out?

Stoney: The point I was trying to make about stability is that a natural, undisturbed forest has to be stable if you're going to protect certain species. You can't just say, we're going to protect this plant or this animal, and put a reserve around it. You have to protect the whole habitat. You have to take an ecosystems approach. The problem is we don't know how big those reserves have to be for a lot of different kinds of biodiversity because we don't even know what the species are that we want to protect for the future.

Agroforestry does provide a way to manage land in these margins between the protected areas and the agricultural areas. In Indonesia, part of the problem is that people want to clear the forestland and actually convert it to farmland. Agroforestry, I think, can help provide some of what they need both from agriculture and from forestry and can create an intermediate land use in between the two areas.

Paarlberg: On the question of deforestation, I suspect the economists and the forestry specialists alike are making the problem seem more difficult than it is. Economists and foresters are interested in the use of trees and in the difficult trade-offs between different uses for fuel, building materials, lumber, and watershed protection. Where deforestation is progressing most rapidly, however, is in Latin America, where the trees being cut down are often not used at all. They are being destroyed by land speculators to convert forest areas to nonsustainable agricultural uses.

Much of our discussion here has been centered on Asia, where the trade-offs are difficult and acute. That engages the economist and foresters in a fascinating dialogue over how to resolve the trade-offs. But the problem of deforestation is accelerating in a part of the world where most of the felled trees aren't being used—they are simply being destroyed.

Bentley: One solution would be to make those trees more valuable by improved technology for utilizing them. People tend not to burn stuff that's worth something.

Berg: This is a short comment complementing the biological diversity issue with the cultural diversity issue because forests around the world are where we find more cultural diversity and more cultural challenges than many other places. I've been sitting in part-time for about 15 months in part of the World Bank dealing with industry and energy, working on cultural issues. Ninety percent of the people there think that if you can reduce it to a number, then it's real, otherwise it isn't. This is one of these issues that is hard to reduce to a number. There are about 10 percent of the people who realize that it is a problem, but they are conflicted over it because their loyalty at the World Bank is to the nation-state. If they start worrying about the subdivisions of the nation-state, particularly these people's, then they may, in their minds, help destroy the nation-state because you have to get so culture-specific that you weaken adherence to a nice uniform national plan. So, the best thing to do is to have a national program that no culture can really fit terribly well, but it is loyal to your client.

One of the real challenges in the World Bank, which is going to become predominant in development financing, is to help find ways in which you can package major money but be culture specific, be it for people or for biologies. It's one of the issues that we're going to have to help them through because they're not able to help themselves to do it.

Kellogg: I think it is true that we really don't know the value of the biodiversity we have. We need to do more research and thinking about the value of biodiversity. But my sense is that we will always have a large amount of uncertainty about that.

When you have that kind of uncertainty, the strategy you ought to use is to think about the consequences of making errors. Type one error results if it turns out biodiversity was very important and we destroy it. What does that cost to correct? Type two error results if it turns out it's really not important and we preserve it. What does that cost?

My sense is that the cost of the type two error, that is, that it's not important and we preserve it, is going to be a lot less than the cost of the type one error, it is important and we destroy it. And, on that basis, we do need to be serious about that preservation. This leads me to say, there are not enough policemen in the world and not enough fences to keep people out of the forests if they need the products that it provides for their livelihood. I think we've learned that over time. So, you must provide a way for people to get access to those minor and major products of the forests that they're going to demand and they're going to have one way or another.

Farm and community forestry programs allow us to start to look at ways of providing increased capacity for small holders to provide those kinds of services and goods in smaller woodlots. That gets us into encouraging broader base rural development and equity. So I believe that farm and community forestry not only points us toward an important strategy for rural development, but also points us toward preserving the biodiversity that may well be very important.

6
Range and Wildlife Resources

Introduction

FEE BUSBY
Winrock International

I've noted that, throughout the symposium, issues associated with rangelands have been mentioned in virtually every talk. Bob Havener, in his opening presentation, talked about the Dust Bowl of the 1930s and what were or were not the causes of it. I don't remember the 1930s, but I do remember the droughts and associated dust bowls of the 1950s and 1970s. They originated from rangeland in the western U.S. that had been converted from permanent rangeland vegetative cover into plowed fields.

Pat Peterson talked about learning to manage natural resources, not protect them. Obviously, that means we should learn not to abuse them. Pat also said we need to develop expertise in managing agroecological systems. Having spent 25 years of my career working in rangeland agroecological systems of the western U.S., I believe that the experiences—both positive and negative—with rangeland management can help develop better approaches to agroecological systems.

In response to Neva Goodwin's question to Dick Harwood about the potential impact on soils from harvesting crops for biomass energy, I would point out that many crops that are being considered for biomass energy are herbaceous rangeland perennials. Most of the organic matter in these plants is in the deep root systems, not in the aboveground plant material that would be harvested for energy.

Rob Paarlberg said that people are more likely to overutilize a resource when they believe their future use is threatened, and that's when the real damage occurs. This behavior is not because people are poor, but because their future use is threatened, and they want to get what they can now. Rob's line of argument reminds me of a number of rangeland situations, including the question of whether ranchers will be allowed to continue to graze livestock on the public lands in the U.S. How does this uncertainty of continued use influence ranchers' behavior?

David Seckler's presentation on the allocation of irrigation water to farmers or other users reminds me that much of this disputed water supply originates on rangeland ecosystems located in the upper part of the watershed. Are there ways to augment the water supply and reduce the conflict through improved rangeland management?

Ecosystem Dynamics and Economic Development of African Rangelands: Theory, Ideology, Events, and Policy

JIM ELLIS
Winrock International

Arid and semi-arid lands cover about one-third of the earth's land mass but nearly two-thirds of the African continent. The majority of African livestock and possibly 30 million livestock-dependent people reside in these dry zones along with the greatest and most diverse concentrations of large wild mammals in existence. The welfare of these people, livestock, and wildlife and the sustainability of Africa's rangeland ecosystems are a source of some concern: Two-thirds of the world's most disadvantaged countries are found in Africa's arid and semi-arid zones (UNSO 1992). However for a variety of reasons, international support for the economic development of these disadvantaged drylands has declined over the last decade. It has, to some extent, been replaced by other forms of international intervention including military missions, famine relief, and desertification control. This shift away from strategic development toward a regime of crisis management reflects the prevailing attitude in development agencies that little can be accomplished in African rangelands, which appear to be overwhelmed by drought, degradation, and political chaos. So the field has largely been left to donors who specialize in disaster relief and community-based help.

There are several reasons for the departure of development agencies from African rangelands. The most often cited cause is the very poor record of return on investment for range and livestock projects in Africa. The poor returns resulted not from the ignorance and intransigence of pastoralists but from a series of misconceptions about rangelands including inadequate scientific knowledge about the dynamics of dry tropical ecosystems, a lack of appreciation of African climate patterns and socioeconomic conditions, and the application of inappropriate range management practices based on theory developed for North American rangelands (Grandin 1987, de Haan 1990).

Another important reason for the loss of interest by development agencies is the image of pastoralists and livestock that has evolved in this century. The public in North America and Europe as well as development planners and government policy makers view people who raise livestock as "herders from hell"—selfish, ignorant folk who have no knowledge or appreciation of the environment around them and who are bent on amassing the maximum number of livestock possible.

Natural events like droughts also shape opinions and policies. Although climate studies show drought cycles to be a natural and recurrent part of the African environment, they are beyond the experience of most expatriate development specialists. They, unlike indigenous pastoralists, have a difficult time incorporating the effects of a lengthy and serious drought into development plans and strategies. It is hard to provide appropriate technical assistance inputs to a system that keeps changing.

Thus pessimism about pastoralists and livestock derives both from development failures and from myths and ideology built up over the past half century about pastoral people, African livestock, and overgrazing. These myths, inappropriate theories, and other forms of misconception, as much as any difficulties inherent in African socio-political systems, have stymied past development efforts and resulted in the current policy of crisis intervention. But this needn't be the case. Results from African climate and ecosystem research carried out in the 1980s have provided new information upon which improved development perspectives and policy may be founded (Behnke, Scoones, and Kerven 1993). We are now in a position to reactivate efforts to improve human welfare and foster environmental conservation in African rangelands. But first, rangeland development policy must be redesigned to accommodate the knowledge gained in the last several years and to eliminate attitudes and interventions based on inappropriate theory, myth, and ignorance.

Equilibrium Theory and Range Management

Nearly a decade ago, Stoddard, Smith, and Box (1975) pointed out that no new conceptual framework had emerged in the field of range management since 30 years earlier when the first edition of their book *Range Management* was published. They might as well have said that the conceptual framework of range science had really not changed since the turn of the century when Cowles (1899) and Clements (1916) developed the theory of ecological succession. Clementsian succession is an equilibrial concept developed in the mesic grasslands of North America. It is based on observations that ecosystems exist in a state of equilibrium with regional climate patterns and that disturbances to the system will result in an orderly and predictable succession back to the equilibrium state, the climatic "climax." It was upon this theoretical basis that North American range management practice, centered on the estimation of carrying capacities and on range condition and trend analysis, was built (Sampson 1923).

In 1969 Eugene Odum (1969) placed the capstone on 70 years of successional research and theory when he proposed that modern ecology be interpreted and ecological crises ameliorated in a successional context. However in the 1970s, even as Stoddard and his colleagues were extolling the immutability of the equilibrial paradigm in range management (Stoddard, Smith, and Box 1975), other ecologists were busy undermining successional theory by documenting the nonequilibrial dynamics and unstable characteristics of many ecosystems and emphasizing the random nature of ecological change (Holling 1973, Noy-Meir 1975, Wiens 1977). But these nonequilibrial perspectives failed to penetrate range science during the 1970s. African range management and livestock development policy in the 1970s and 1980s were thus based on equilibrial premises transplanted from North America; the results were generally dismal, causing agencies like USAID and the World Bank to give up on range and livestock projects. It wasn't until the late 1980s that the new nonequilibrial ecological concepts began to be explored in range science when climatically induced crises in Africa and Australia caused ecologists to question the applicability of the North American equilibrial paradigm in highly unstable, dry tropical ecosystems (Caughley, Sheperd, and Short 1987; Ellis and Swift 1988; Westoby, Walker, and Noy-Meir 1989). To date, these new perspectives have had no discernible effect on African range and livestock policy despite several years of discussion (for example see summaries in Cincotta, Gay, and Perrier 1991 and Behnke and Scoones 1991).

Myth and Ideology: The Mainstream View

Pessimism about African rangeland development and the reputation of livestock and pastoralists as agents of environmental destruction grew rapidly during the 1970s and 1980s. Sandford (1983) called this pessimistic perspective the mainstream view because it was widely accepted. One cause of the pessimism was the near-universal failure of donor-sponsored livestock development projects. This was understandably discouraging but largely predictable. Livestock projects usually had as prominent goals the confinement of people and animals to group ranches or grazing blocks and the reduction of livestock numbers. Thus pastoralists' reluctance to settle, destock, and go into market-oriented beef production schemes predestined the failure of many projects. But the mainstream view of pastoralism is more than a response to the failure of development projects; it is an ideology, growing from myths about pastoral tradition, pastoral land tenure and land-use practices, and the effects of grazing on dry ecosystems.

Perhaps the formative element of the mainstream ideology is the myth of pastoral irrationality (Herskovitz 1926), i.e., the idea that pastoral tradition requires amassing large numbers of livestock solely for reasons of prestige and social status. This portrayal of pastoralists as inflexible traditionalists is still part of the conventional wisdom, but even strong critics of pastoralists who have observed them firsthand, allow that they are quite pragmatic and adaptable, rather than irrational (Brown 1971, Lamprey 1983). The most pervasive and prominent aspect of the mainstream ideology, the tragedy of the commons concept, relates to land tenure and range utilization patterns (Hardin 1968). Hardin's construct assumes open-access resource utilization in pastoral systems and concludes that overgrazing must necessarily follow this land tenure arrangement. Although this open-access assumption has repeatedly been shown inapplicable in traditional pastoral land tenure systems, and therefore a myth with respect to pastoralism, it was most influential in the formulation of policies for African rangeland management (Moris 1988, Vedeld 1993).

Desertification provided the linchpin of the mainstream ideology. Since early in the century, people have speculated that the Sahara was expanding rapidly to the south, overwhelming villages, fields, and grazing lands in the process. Sand dune expansion or desert encroachment was attributed to climate change, which in turn was thought to result from overgrazing, tree cutting, and shifting cultivation (Bovill 1921, Stebbing 1935, Dregne and Tucker 1993). This view remains prevalent today with only slight modification and was given modern credence by UNEP's desertification-control campaign, by

Lamprey's estimate that the Sahara was migrating south at a rate of 6 kilometers per year (Lamprey 1975), and by Charney, Stone, and Quirk (1975) who proposed a feasible process whereby vegetation removal and the subsequent expansion of bare soil could feed back to the atmosphere, cause a reduction in rainfall, and lead to desertification.

The mainstream ideology gained strength throughout the 1980s supported by reports of livestock devastating rainforests in South America and so on. The popular press embraced the mainstream ideology so that the informed public and policy makers in North America and Europe now believe that livestock are responsible for worldwide environmental destruction. This public perception, coupled with the failure of African livestock development has had a major effect on development policy. Development agencies and donor-funded research organizations have had little involvement in African rangeland, livestock, and pastoral development activities since the mid-1980s (ILCA 1987).

Natural Events: The Droughts of the Mid-Twentieth Century

Major shifts in climate, from relatively wet periods to dry periods or droughts, are normal in the rangelands of Africa. Long-term rainfall records and other evidence show that these sorts of climate dynamics have prevailed there for at least the last 10,000 years (Nicholson 1983, Hulme 1990). It is this dynamic variability that renders equilibrium theory and North American range management practice inappropriate in many parts of Africa. African livestock herders understand this variability and are thus reluctant to subscribe to projects sponsored by development agencies that ask them to forego their adaptations to climate change and to adopt practices designed for the much more stable ecosystems of North America. But the mainstream ideology, based as it is on human-induced desertification and degradation, would probably not have achieved its great acceptance if the Sahel had been wet over last 50 years, rather than undergoing an extended drought.

The drought in the Sahel probably started in the mid-1950s with a subtle downturn in rainfall following 20 years of above-average precipitation. Forty years later, rainfall levels still have not returned to the long-term mean values throughout most of the Sahel, although the lowest precipitation levels may have been reached in the mid-1980s. During this period, serious droughts have also occurred in eastern and southern Africa, but they have not equaled the Sahelian drought in duration.

Although the rainfall decline began in the 1950s, it was not until the mid-1970s that the scale and seriousness of the Sahelian drought became apparent. Even then, the history of drought in Africa was not well appreciated, and causes for this extended drought were not well understood, but the finger of suspicion pointed at people and livestock. Overgrazing and overexploitation of arid and semiarid lands is a age-old concern that comes to the fore whenever drought strikes. The long-delayed passage of federal legislation regulating livestock grazing in the United States was probably aided by the drought of the 1930s, which devastated North American drylands. In a similar way, the Sahelian drought resurrected concerns voiced early in the century about the role of African livestock, pastoralists, and other traditional people in range degradation, local desertification, and the advance of the Sahara to the south (Lamprey 1975; Charney, Stone, and Quirk 1975; Sinclair and Fryxell 1985). In essence, the ecological changes accompanying the Sahelian drought were perceived in a equilibrial context. Human overexploitation was believed to have disrupted ecosystem equilibrium, leading to or at least exacerbating the drought. The UNEP desertification-control program was a policy response designed to meet this crisis through ecosystem rehabilitation and restoration of equilibrial conditions.

Thus, during the decades of the 1970s and 1980s, arid and semi-arid lands in Africa were perceived to be under severe threat from drought and from the inhabitants of these regions, who were portrayed as instigators rather than victims of the ecological crisis (Lamprey 1983). Development policy therefore moved away from rangeland development and began focusing on anti-desertification activities. This policy perspective remains in force today. A new $450 billion desertification-control initiative, based on the mainstream ideology, has recently been launched by UNEP to expand anti-desertification activities in Africa (Pearce 1992).

The 1980s: New Research and the New Paradigms

Climate analyses

While donor-sponsored development in the drylands waned in the early 1980s, basic and applied research did not. The drought in the Sahel and droughts in eastern Africa stimulated a stream of new analyses of long-term African climate patterns. The work of Sharon Nicholson (1983), among others, suggested that the Sahelian drought was typical of long-term climate fluctuations in western Africa. Similar aridification periods have occurred several times within the last 10,000 years. Other studies, focusing on eastern Africa pointed out that although severe droughts are also an integral part of the prevailing climate there, the long-term dynamics of drought are out of phase with those in western Africa and differ in frequency and duration. This led to the suggestion that dealing with drought is a fundamentally different process in different parts of Africa, due to divergent regional climate patterns (Farmer 1986). These studies demonstrated that while the current droughts were severe, they were not unprecedented and thus not solely attributable to modern rangeland exploitation and livestock-induced climate change. This obviously implies that if shifts in desert boundaries occur, they could be a natural outcome of long-term climate fluctuations, rather the result of mismanagement by indigenous people.

Ecosystem dynamics

Partly due to questions about the drought and the assumed degradation of African rangelands and partly due to concerns about the plight of pastoralists, several long-term interdisciplinary studies of livestock-dominated ecosystems began in the early 1980s. Some of these were directly related to desertification-control initiatives and designed to develop procedures and policies for better livestock management and relief from perceived livestock-induced degradation (the UNESCO Integrated Project for Arid Lands). However most projects, including the ILCA systems studies, the U.S. National Science Foundation South Turkana Ecosystem Project, and investigations of communal rangelands in southern Africa (e.g., Scoones 1990) were designed to answer questions about the fundamental relationships between people, livestock, and the environment, rather than assuming desertification as a basis for the research.

Also in the early 1980s, two important publications by authors with long experience in African rangelands established the opposing positions for the debate about the role of livestock in environmental degradation. In "Pastoralism Yesterday and Today," the ecologist Hugh Lamprey (1983) took the position that pastoral livestock tend to degrade their environments and to destabilize equilibrial ecosystems by overgrazing. Formerly, he suggested, this degradation was of minimal concern because pastoralists could migrate from degraded to unexploited environments while the overgrazed regions recovered—a sort of sustainable shifting pastoralism. Today, traditional grazing patterns allegedly lead to desertification because pastoralists no longer can practice this shifting pastoralism.

A very different perspective was developed by the economist Stephen Sandford (1983) who, in analyzing pastoral development, demonstrated how the opportunistic and risk-aversive practices of traditional African pastoralists were appropriate in highly variable environments. He showed that these nonequilibrial ecosystems are better exploited by flexibility and opportunism rather than by adhering to the conservative strategies associated with North American range management practices.

To make a decade-long research story very short, the results of the various systems studies have, on balance, tended to support the views of Sandford rather than those proposed by Lamprey. While some studies did document limited degradation due to livestock grazing, the more general conclusions supported the concept that traditional African livestock husbandry practices were compatible with Africa's climatically driven, dramatically unstable, nonequilibrial rangeland ecosystems and were not inherently destructive. These and a number of similar studies carried out simultaneously in Australia, produced a series of papers that, in essence, show that tropical rangeland ecosystems function very differently than those temperate ecosystems where succession theory and modern range management practices were developed. The African and Australian studies found that climate variability plays a much larger role in these tropical systems, which thus require different and more opportunistic management practices than is appropriate in temperate rangelands. Together, these studies instigated a paradigm shift in concepts of tropical dryland ecosystem dynamics, in views about their management, and in perceptions of the livestock husbandry practices of indigenous African herders (Ellis and Swift 1988; Westoby, Walker, and Noy-Meir 1989; Mentis et al. 1989).

Remote sensing of desert dynamics

While climate analyses and interdisciplinary ecosystem studies were clarifying the dynamics of African rangelands, a new remote-sensing technology was being developed that severely undercut the perception of rampant degradation and the spread of the Sahara, i.e., the concept that anthropogenic desertification was under way in Africa. In late 1981, a U.S. meteorological satellite began transmitting small-scale imagery of the planet on a twice daily basis. Although the satellite was designed to study weather, C. J. Tucker and his colleagues at NASA began using its frequent and inexpensive imagery to study the dynamics of the earth's vegetation. One of several applications of this NDVI (normalized difference vegetation index) data was a continental-scale analysis of the long-term movement of the boundary between the Sahara and the Sahel. Based on a decade of data collection and analysis, Tucker and his colleagues found no evidence that the Sahara was advancing to the south (Tucker, Dregne, and Newcomb 1991). They did find that the Sahel-Sahara boundary is very dynamic and moves both north and south in response to annual changes in rainfall, but no long-term directional trend is evident.

Hellden (1988) also used long-term remote-sensed vegetation data and historical aerial photography to re-examine some of the original evidence used to validate desertification. Surveys conducted in the mid-1970s claimed that vegetation had disappeared and sand dunes were advancing on villages in central Sudan where no dunes existed before. These surveys provided strong justification for the need to reduce livestock numbers and to develop the multi-billion dollar UNEP Desertification Control Program. However Hellden's exhaustive analysis showed no change in sand-dune prevalence and little change in vegetation in those villages initially used to document the desertification process.

These results from long-term remote sensing studies not only weaken the desertification concept, they also support the findings of climate analyses and ecosystem studies that many of the drylands of Africa are indeed nonequilibrial and unstable environments in which vegetation abundance waxes and wanes dramatically with annual rainfall patterns. Nevertheless, there remains little evidence of widespread or long-term degradation or desertification in sub-Saharan Africa, and there is abundant evidence that the strategies of pastoral peoples are indeed compatible with the dynamics and structure of the ecosystems they inhabit.

Policy results

The research conducted in the 1980s generated a consensus among many scientists that desertification is not an overwhelming problem in most of sub-Saharan Africa (Nelson 1988), that African livestock enterprises do not necessarily degrade the environment (Mace 1991), and that new development approaches are needed to deal with the unique climate-environment interactions prevalent in sub-Saharan rangelands (Behnke, Scoones, and Kerven 1993). But this consensus among researchers has had little or no impact on development policy despite the fact that it has been several years since the relevant evidence began mounting. In fact, livestock continue to be seen as a major threat by the public (Durning and Brough 1991) and rangelands remain a non-priority on the agenda of development agencies and donors. For example, in the late 1980s, ILCA was directed by its donors to curtail its innovative and productive pastoral systems studies, to reduce work in the dry zones, and to adopt a commodity focus (ILCA 1987). The Technical Advisory Committee to the CGIAR centers reiterated in 1993 that livestock work in the international agricultural centers is overfunded; both ILCA and ILRAD recently suffered substantial cuts in their research budgets.

Why have the research results of the 1980s and the new nonequilibrial paradigm had no discernible effect on policy? Part of the problem is that researchers have not successfully communicated their conclusions to the donors and the public despite the publication of numerous scientific papers. There is usually a long time-lag between the generation of new knowledge and the absorption of that knowledge by the public and policy communities. It is also likely that scientists are naive about the policy process and not particularly effective at getting a hearing for their results and opinions. Efforts are now under way to translate the research results of the 1980s and the new paradigm into practical management procedures (Behnke and Scoones 1992). Nevertheless rangeland development policy remains dormant in Africa. In contrast, the issue of desertification remains on the policy agenda and may be experiencing a resurgence despite evidence that the problem has been extravagantly exaggerated (Pearce 1992). Concern has been expressed that a massive new desertification-control program will capture all available funds and eliminate other forms of development and environmental conservation in African rangelands. In other words, the new desertification-control initiative, if successful, may help keep rangeland development off the policy agenda.

Where to Next?

There are good reasons why desertification control should not be the major form of development intervention in the arid and semi-arid zones. Most obviously, the weight of scientific evidence shows that human-induced desertification is not an overwhelming problem. In contrast, there are serious economic, human welfare, and related political problems that do need attention. Some of Africa's dry-zone countries are among the world's most economically disadvantaged, and some are quite unstable politically (UNSO 1992). The scientific progress of the last decade provides us with a knowledge base upon which we can begin to formulate new development approaches to advance human welfare, maintain environmental values, and work toward economic growth and political stability. However new initiatives must take account of the fact that climate variability, drought, and aridification are an integral part of the dynamics of African drylands.

Stable agricultural production has been suggested to be a prerequisite for economic development, and so it has been in Europe, North America, and Asia (Winrock 1991). But it is hard to see how agricultural production stability can be achieved in dry Africa where rainfall, the primary driver of production, is highly unstable. This suggests either that rainfed agriculture and livestock husbandry may not be appropriate bases for development in African drylands or that we need to generate a new Africa-specific model for economic development. The models and methods used in Asia and the temperate zones may be useless here. Fabricating a new approach to development for Africa may be a difficult process, but as a first step perhaps we should acknowledge that indigenous pastoralists have developed methods for dealing with variability that have worked reasonably well for at least 3,000 years.

Literature Cited

Behnke, R. H., and I. Scoones. 1992. *Rethinking range ecology: Implications for rangeland management in Africa*. Issues Paper no. 33. London: International Institute for Environment and Development.

Behnke, R. H., I. Scoones, and C. Kerven. 1993. *Range ecology at disequilibrium: New models of natural variability and pastoral adaptation in African savannas*. London: Overseas Development Institute.

Bovill, E. W. 1921. The encroachment of the Sahara on the Sudan. *J. Afr. Soc.* 20:174-85, 259-69.

Brown, L. H. 1971. The biology of pastoral man as a factor in conservation. *Biological Conservation* 3(2): 93-100.

Caughley, G., N. Sheperd, and J. Short, eds. 1987. *Kangaroos: Their ecology and management in the sheep rangelands of Australia*. New York: Cambridge University Press.

Charney, J. G., P. H. Stone, and W. J. Quirk. 1975. Drought in the Sahara: A biophysical feedback mechanism. *Science* 187:434-35.

Cincotta, R. P., C. W. Gay, and G. K. Perrier. 1991. New concepts in international rangeland development: Theories and applications. In *Proceedings of the 1991 International Rangeland Development Symposium*. Logan: Department of Range Science, Utah State University.

Clements, F. E. 1916. *Plant succession: An analysis of the development of vegetation*. Publication 242. Washington: Carnegie Institution.

Cowles, H. C. 1899. The ecological relations of the vegetation on the sand dunes of Lake Michigan. *Botanical Gazette* 27:97-117, 167-202, 281-308, 361-91.

de Haan, C. 1990. Changing trends in the World Bank's lending programme for rangeland development. In *Suitable yield system: Implications for the world's rangelands*, ed. R. Cincotta. Logan: Department of Range Science, Utah State University.

Dregne, H. E., and C. J. Tucker. 1993. *Satellite monitoring of desert expansion*. In *A global warming forum: Scientific, economic, and legal overview*. Boca Raton, Florida: CRC Press.

Durning, Alan B., and Holly B. Brough. 1991. *Taking stock: Animal farming and the environment*. Worldwatch Paper 103. Washington, D.C.: Worldwatch Institute.

Ellis, J. E., and D. M. Swift. 1988. Stability of African pastoral ecosystems: Alternate paradigms and implications for development. *Journal of Range Management* 41:450-59.

Farmer, G., 1986. Rainfall variability in tropical Africa: Some implications for policy. *Land Use Policy* 3(Oct.): 346-52.

Grandin, B., 1987. Pastoral culture and range management: Recent lessons from Maasailand. *ILCA Bulletin*, no. 28:7-13.

Hardin, G., 1968. The tragedy of the commons. *Science* 162:1243-48.

Hellden, U. 1988. Desertification monitoring: Is the desert encroaching? *Desertification Control Bulletin* (UNEP, Nairobi), no. 17:8-12.

Herskovitz, M. J. 1926. The cattle complex in East Africa. *American Anthropologist* 28:230-72, 361-80, 633-64.

Holling, C. S. 1973. Resilience and stability of ecological systems. *Annual Review of Ecology and Systematics* 4:1-23.

Hulme, M. 1990. The changing rainfall resources of Sudan. *Transactions, Institute of British Geographers* 15:21-34.

ILCA (International Livestock Centre for Africa). 1987. *ILCA's strategy and long-term plan: A summary*. Addis Ababa, Ethiopia.

Lamprey, H. F. 1975. Report on the desert encroachment reconnaissance in Northern Sudan. 21 Oct. to 10 Nov. 1975. UNESCO/UNEP. Mimeo.

Lamprey, H. F. 1983. Pastoralism yesterday and today: The overgrazing problem. In *Ecosystems of the world*. Vol. 13, *Tropical savannas*, ed. F. Bourliere, 643-66. Amsterdam: Elsevier.

Mace, R. 1991. Conservation biology: Overgrazing overstated. *Nature* 349:280-81.

Mentis, M. T., D. Grossman, M. B. Hardy, T. G. O'Connor, and P. J. O'Reagain. 1989. Paradigm shifts in South African range science, management and administration. *South African Journal of Science* 85:684-87.

Moris, J. 1988. Failing to cope with drought: The plight of Africa's ex-pastoralists. *Development Policy Review* 6:269-94. London and New Delhi.

Nelson, R. 1988. *Dryland management: The desertification problem*. Working Paper No. 8 for ENVST. Washington, D.C.: World Bank.

Nicholson, S. E. 1983. Climate variations in the Sahel and other African regions during the past five centuries. *Journal of Arid Environments* 1:3-24.

Noy-Meir, I. 1975. Stability of grazing systems: An application of predator-prey graphs. *Journal of Ecology* 63:459-81.

Odum, E. P. 1969. The strategy of ecosystem development. *Science* 164:262-70.

Pearce, R. 1992. Mirage of the shifting sands. *New Scientist* 136:38-42.

Sampson, A. W. 1923. *Range and pasture management*. New York: John Wiley and Sons.

Sandford, S. 1983. *Management of pastoral development in the Third World*. New York: John Wiley and Sons.

Scoones, I. 1990. Livestock populations and the household economy: A case study from southern Zimbabwe. Ph.D. diss., University of London.

Sinclair, A. R. E., and J. M. Fryxell. 1985. The Sahel of Africa: Ecology of a disaster. *Canadian Journal of Zoology* 63:987-94.

Stebbing, E. P. 1935. The encroaching Sahara. *Geological Journal* 86:510.

Stoddard, L. A., A. D. Smith, and T. W. Box. 1975. *Range management*. New York: McGraw-Hill.

Tucker, C. J., H. E. Dregne, and W. W. Newcomb. 1991. Expansion and contraction of the Sahara Desert from 1980 to 1990. *Science* 253:299-301.

UNSO (United Nations Sudano-Sahelian Office). 1992. Assessment of desertification and drought in the Sudano-Sahelian Region. New York.

Vedeld, T. 1993. *Environmentalism and science—Theory change on collective management of natural resource scarcity*. Working Papers, vol. 1, no. 1. Aas, Norway: Noragric.

Westoby, M., B. H. Walker, and I. Noy-Meir. 1989. Opportunistic management for rangelands not at equilibrium. *Journal of Range Management* 42:266-74.

Wiens, J. A. 1977. On competition and variable environments. *American Scientist* 65:590-97.

Winrock. 1991. *African development: Lessons from Asia*. Arlington, Virginia.

Panel

JIM MANER
Winrock International

The drylands and the semi-arid areas of Latin America are quite different from those of Africa. I would like to take the opportunity to point out one or two differences, looking specifically at Mexico. Much of Mexico is semi-arid. Most of the drylands are agro-pastoral, meaning that sedentary groups of populations are utilizing them. Historically, through the political system and the revolution that brought about agricultural reform in Mexico, many people who had little background in agriculture were settled in the dryland areas and were expected to make a living.

In a project in Mexico, we've been working with *ejidos*, or communal lands, which might have a land area as large as 14,000 hectares with about 140 to 150 families. They attempt to produce maize and beans when there's enough moisture. But due to low levels of and tremendous variation in rainfall, they don't always produce maize and beans. Most of them depend heavily on livestock. It is an economic imperative that they produce livestock—sometimes cattle, but mainly goats that produce milk and kids that they sell. They depend on these livestock for their income, therefore, they utilize them to the greatest extent possible.

I won't agree that degradation does not occur from overgrazing by livestock. I think we can document that it does occur. If you increase the level of livestock in certain areas, you can definitely show overgrazing and destruction. In Mexico if we look at water points or water sources and move outwards you will notice degradation close to the water points. And by degradation, I mean elimination of almost all ground cover except for shrubs.

In most Mexican drylands, as well as in much of the U.S. rangelands, there has been a tremendous increase in shrub production—as high as 65 to 70 percent of the plants on drylands are shrubs—and a reduction of grasses and forbs, which are the high-value forage for most livestock. Recent data shows that this occurs whether or not we have livestock grazing on it or not. We know that if we leave these areas without livestock, there will be an increase in shrub production, but also there is an under-story growth of grasses and forbs that protects the soil and reduces soil erosion.

In Mexico grazing causes tremendous destruction of the under-story. Basically, we're left with the shrubs, which are of lower value for animals and do not protect the soil, so increase soil destruction increases. The soil is degraded in such a way that it cannot replenish itself in the areas that we're talking about.

One of the problems is institutional. Because of the way that the ejidos are set up, there is no limitation on the number of livestock that each of the farmers can place on the ejido land. Livestock, as in Africa in many cases, is wealth and the more animals you have, the greater income that you might derive from them. As a result, each farmer is trying to increase the number of goats. However, increasing goat numbers doesn't always increase their production. Because of the competition for limited forage that is available, actual production goes down and economic income decreases as well.

Range management specialists talk about grazing management and grazing pressure. They talk about eliminating or reducing the number of animals. We found that even reduction in animal pressure within certain ranges does not have much effect on the actual level of shrubs and level of grasses that are available.

So we need to do more than grazing management in these areas. Farmers depend on livestock. Livestock continue to utilize the dryland shrubs and the dryland forage that are available and eke out an existence. We have to think about what policies might change how farmers make their living. I think we need to look at alternative enterprises and alternative ways for these people to utilize the ranges not only for animals but for other things.

NED RAUN
Winrock International

The drylands—the arid and semi-arid agroecological zones—are the toughest problem you can get your hands on because of the water constraint.

The water is either there or it comes from the clouds or it doesn't, and you don't have any alternative. That's why you struggle with it. That's why ILCA struggles with it. That's why the CGIAR struggles with it. In the higher rainfall areas, you have constraints, but you also have some options on what you might do.

The *Assessment of Animal Agriculture in sub-Saharan Africa* [Winrock International Morrilton, Arkansas, 1992] estimated potential capacity by agroecological zone. The current stocking rates are estimated to range from 30 to 100 percent. The driest areas are probably at capacity—there's little opportunity to bring in more animals over time. As you go south from the Sahara and the rainfall increases, there may be an opportunity to increase the number of animals. In arid and semi-arid areas, which account for about 60 percent of the land area of sub-Saharan Africa, you find about 55 percent of the ruminant animals. The subhumid area, which accounts for 20 percent of the land area of sub-Saharan Africa, has about 20 percent of the ruminant animals. It is now considered to be at about 30 percent of capacity. So there's an opportunity here to triple livestock numbers.

In a background paper on desertification done for the *Assessment of Animal Agriculture*, Jerry Dodd, from the University of Wyoming, made the point that there's a tendency for casual observers to exaggerate the range condition when they look at it in the dry season. In the arid region, rainfall is low, and the vegetation is dominated by annuals. Any farmer or herdsman knows that, when you deal with annuals, in the rainy season you have lush growth, and when the dry season comes, the annuals crumble and practically disappear. But when the rains return, the seed that's in the ground comes forth and it's green again. That's why the casual observer needs to be careful about making judgments on range conditions seen in the dry season.

Another point: I happen to have lived through the dust storms in the 1930s. They came about for two reasons. Climate was one, but the big factor was bad farming practices—practices where farmers left no cover on the land. When it didn't rain, the winds came and we had dust storms. There's a lesson there.

In the Sahel, in northern Senegal as an example, when they harvest the peanuts, they remove all of the vegetation from the land, and leave it bare. That practice is not sustainable. So much of the so-called severe degradation that you see in the Sahel is in consequence of inappropriate farming practices. They attempt to make some of these dry rangelands into cropland, but these lands are not suitable for cultivation.

The emphasis in land management for the arid and semi-arid lands needs to be on stabilization of those systems. Jim Ellis mentioned opportunistic range management, that is, management driven by climate and multispecies grazing. But a third factor that's important in these areas is the empowerment of the local people to govern themselves, to control the grazing of their land, the water points, and the services that are provided to the community.

When development assistance began, pastoralists did have control over their land and their grazing system and the water, but unfortunately development specialists then came along and decided that they needed technology where there were more controls. It has been a big flop. Now the development community is going back to where we were in the beginning, that is, looking for local empowerment, local control, and, then, insofar as possible, the privatization of services instead of having the government providing them.

In the subhumid zone, I think the emphasis must be on looking to these lands on how production can be increased. If it's going to be increased, production systems will have to be intensified.

I will conclude with a few comments on arable land. In sub-Saharan African, about 85 million hectares are now under cultivation in food crops. If you add nonfood crops, it's something over 100 million hectares. Looking ahead to 2025, to meet the needs for food for direct human consumption, along with some feedgrain for poultry and pigs, it will be necessary to double the cultivated land area to 170 million hectares. It has been estimated that up to 800 million hectares of land may be potentially cultivable. However, much of that land cannot be practically developed for economic and environmental reasons. If we're going to meet future needs for food to feed people and for income and employment, crop yields and animal productivity will have to increase.

The arid lands will continue to have livestock because that is the only way to harvest a product that is useful to man. But looking ahead to 2025, our estimate is that only about 45 percent of the ruminant population in sub-Saharan Africa would be there. It's now 55 percent.

FANNY NYARIBO-ROBERTS
Washington State University

Jim Ellis contrasted two views of the role of livestock in African ecosystems and how livestock affect the environment. The first is that livestock continue to be seen as destroyers of the environment through overgrazing and clearing of forestlands, which in turn have disrupted the ecosystem equilibrium. The second view is that the opportunistic management practices of African pastoralists are appropriate largely due to the highly variable environment in which these systems operate. Therefore, African livestock production practices are compatible with the climatically driven and unstable ecosystems. I would like to add another dimension. It is that of a diminishing land base, in the high-density, high-potential areas where agricultural production is of a sedentary agro-pastoral nature.

Jim Ellis' paper points out that based on research results in the last 5 years, there's a consensus among scientists that desertification is not a problem in most of sub-Saharan Africa, that African livestock enterprises do not necessarily degrade the environment, and that there's a need to deal with the unique climate-environment interactions. Apparently results from this research have not been internalized by policy makers and funders of research.

Whatever view is held by policy makers, researchers, and donors, the question is, what is to be done about the existing ecosystem disequilibrium, whether man-made or due to climatic reasons? In my opinion, crop-livestock integration will continue to be important and should continue to be stressed.

Most observers suggest a systems approach to environmental and natural resource management problems. From a methodological point of view, my fear is that already the systems approach has been criticized as being too narrow, too capital-intensive, and too slow. And there has been a lack of impact, particularly in livestock projects where you have breeding components and disease problems. I believe these factors should be borne in mind in designing future research and development projects.

Discussion

Blake: I'll tell you a cautionary tale. When I was in Mali, I was interested in the problem of getting cattle from the sub-Saharan area of Mali down to Abidjan on the coast in a decent condition so people could have good meat and so the sellers could earn more money. Today, as in the past, the cattle are walked almost a thousand miles to market, so they are skeletons when they reach the coast.

We had the idea of setting up a series of feeding stations where the cattle would eat the cottonseed and other locally available feeds and have dependable water supplies so they would arrive in adequate shape. USAID made a fairly good investment in this project. Before we got going, I insisted that USAID send an anthropologist to talk with people to be sure they supported the project.

So, we went ahead. After spending a lot of money, the project bombed. The reason was that we didn't know how to gauge the real feelings of the Fulani herders—that when they said yes, they really meant no. These Fulani felt quite detached from the Mali government and indeed from other tribal groups. They didn't really want to be part of the national economy. They wanted to keep to themselves and do what they wanted. Their purposes for keeping cattle were to provide a little food, but more important to acquire prestige that is based on the number of cattle they own and to provide bride prices. They simply would not sell most of their cattle. The only ones they would sell were ones that had the evil eye—the cow that kicks over all the buckets and the fences. We were trying to integrate a whole way of tribal life into a modern structure when the people didn't want to do it. The younger Fulani didn't want it any more than the older people.

Ellis: I think that is not untypical for livestock-raising people in highly variable environments. Steven Sanford had this figured out in the early 1980s and we haven't paid any attention to him. He coined the term "opportunistic management," to the best of my knowledge. That's what they're doing. They're keeping their animals as long as conditions are good, and they're ready to unload them when conditions get bad. When they run short of animals, they go out and get them in whatever way that they can. The challenge is to try to figure out how we can help them to opportunistically manage within some sort of system that actually has a marketing and modern economic framework. It's not an easy question. All over Africa we need a vast improvement in marketing systems, and until marketing systems are available that are reasonably flexible, we're not going to be able to alter these systems or modernize them in any respect.

What Roy Behnke has suggested in his work is that there really is no reluctance by pastoralists in general to sell animals if the conditions are right. That's a little counter to what we normally think, but I suspect that he's probably right. Maybe when you were trying to promote this sort of marketing enterprise, the incentives just weren't there at that time for people to sell these animals.

Coward: Is this an arena in which there are opportunities to integrate indigenous knowledge along with the knowledge of scientists and ecologists? I wonder how much we know, for example, about the local institutional arrangements that people have put together to deal with the highly erratic climatic situation in which they're operating?

Another question, can we imagine rural economies in this part of the world, based on these kind of ecological situations, in which people really can have an improved life? Or are we talking about people remaining at a very low subsistence level and not having opportunities for education, health, other things that growing economies can provide?

Ellis: In my opinion it's not morally acceptable to ask the last question. We must aim at providing opportunities for a better lifestyle for these 200 million people.

I have worked in northern Kenya for over a decade now. I'm reminded of a Turkana woman, an older woman, who we, in one of our many interviews, asked about the nomadic lifestyle and what she thought about it. She said she would love to settle down and take life easier, to have a transistor radio, to have a house to come home to instead of migrating around on a regular basis. In essence she said that if they had any viable option, they would much rather have a richer, more comfortable life.

These people do very well under the circumstances. Certainly they appreciate what the Maasai called "the life," as many of us appreciate the value of being out of doors and the things that go along with that. On the other hand, I don't think there's any question they value improvements in their material wealth, in education—all of those opportunities. But, if you're going to live in that kind of environment and depend on it, then the traditional lifestyle is a successful one. So the challenge is how to provide opportunities that people want and need and, at the same time, take advantage of the rich indigenous knowledge that these people have. They know how to live in these difficult environments. I'm not thinking so much of integration of range management knowledge with indigenous knowledge but, rather, range managers learning from indigenous people and then trying to figure out how to provide better opportunities and better lifestyles for these people. It's not going to be easy. We have to start understanding that these people are not destroying their environment; they've been using it for a long time in a relatively conservative way.

Nyaribo-Roberts: Sometimes we talk about low-input sustainable agriculture, but in Africa you have low-input agriculture that is unsustainable. I have worked in areas with high rainfall and good soils but where the soil nutrients are rapidly diminished due to continual use. I think to improve the quality and lifestyles of these people beyond semi-subsistence living, external inputs such as credit and markets will be necessary.

The farmers that I've worked with know how to use residues. They know how to use maize stover. They know how to use maize thinnings. They grow maize right up to the homestead. So they are very efficient users of the resources they have. I think what they need in certain instances is to have other external inputs that are not included in technological packages introduced by research and development projects.

Berg: In the early 1980s, when I was chairing the donor group at the OECD evaluating donor assistance, we ran an exercise comparing official data and the donor results. It was clear that the higher the engineering content, the better the results, and the higher the social content, the worse the results. When it came to pastoral projects, we couldn't find a single success. Jim Ellis said that was true of the 1970s, but the implication is that the record has gotten better in the 1980s. Yet, CGIAR seems to have run counter to that evidence.

Ellis: What I intended to convey was that the ideas and policies that were developed during that period still prevail. I think our knowledge has improved dramatically since those days, and I think ILCA has done some wonderful work on these issues. However, they were told to cease because there wasn't thought to be much future in trying to improve livestock systems and pastoral systems. I didn't mean to imply that, from a development perspective, things have gotten better since the early 1980s, but, from a research and knowledge-generation perspective, I think we know a lot more.

Paarlberg: Haven't pastoral systems in Africa already broken out of some of the traditional limits that made them environmentally sustainable? Years ago, cattle populations were limited in Africa not so much by water availability or rangeland conditions, but by animal disease, tribal rivalries, and cattle raiding. Then with colonialization and in the post-colonial period, veterinary medicine arrived. Colonial administrations and post-colonial administrations also improved population security, so animal numbers doubled and redoubled. The constraint on total populations suddenly became rangeland conditions.

That's fine if you have stable rainfall but, as you noted, you have highly variable rainfall. So, when populations go up to the rangeland limit during good rainfall years, you're looking for serious trouble—casualties in animal numbers and distress in the human population—once you hit a down rainfall cycle. So to what extent can we invoke a traditional system, which functioned well for the environment before veterinary medicine, in the post-veterinary medicine era?

Ellis: I have two sorts of answers. One is that probably in more places than you would think, there isn't any veterinary medicine to speak of. Veterinary services are a big problem in many areas in Africa. They are quite ineffective in some areas. So that's really not a major release from constraints in some areas of the drylands themselves. That's not to say that there aren't veterinary services in some places. In fact, that does raise the question, if disease does not control populations, what are the controls?

Look at the Maasai group ranches in Kenya, which have fairly good veterinary services. The Maasai have been ostensibly confined to relatively small areas on these group ranches, and certainly the livestock populations there have expanded to what one would think is the capacity of those areas. I've got to say that Maasai land south of Nairobi is not an arid environment. It gets dry seasonally, but it's a fairly reasonable environment. However, these people have good market access for their animals. Although their animal numbers have increased, their marketing has increased as well. So in that situation where you do have veterinary services, it is accompanied by better marketing outlets. The places where you don't have markets, you often don't have veterinary services either. It would be a big mistake to develop effective veterinary services for an area where you had no provision for marketing. In fact, I think the first step one has to make in thinking about development in livestock areas is providing marketing opportunities. If you can't create marketing opportunities, you have to think about whether you're going to provide things like veterinary services or not.

Paarlberg: But didn't colonial administrations make that mistake in the Sahel?

Ellis: I think that's right. Let me make one another point, though. Livestock numbers have not increased dramatically in Africa in the last couple of decades. In fact, the last figures I saw, the trend was very slightly upward. So it's not as if they're doubling and redoubling at a constant rate.

Peterson: Livestock in Africa is another indication of how we can take a poor situation and make it worse. It seems to me there are some sort of symbiotic, hopeful systems there that we need to be interested in, if for no other reason than there does seem to be a greater capacity to respond quickly to famines generated by droughts through livestock than through cropping systems because the seasonality of the rain isn't quite as important. I think that that's something that we don't think about when we think about management of livestock, especially in Africa.

Second, I think that we keep putting things in boxes. We talk about forestry, we talk about livestock, and we talk about this, but when you get down to the household level, those things are integrated. We, as the experts, don't seem to be able to integrate them ourselves. We keep getting waves of different expertise coming up to talk about one part of the box, and then they sit down, and we get another wave.

Agricultural research in general has probably been captured too much by plant scientists maybe flowing out of some Green Revolution activities, and that might be part of what's driving the CGIAR. Look at how they're going to allocate their budgets to international centers because budgets are getting tight and people have a hard time thinking broadly.

Finally, I would like to make a comment about desertification. It seems to me that we can't afford to wait 40 years to find out whether or not livestock systems, mismanagement of range, whatever, is causing this serious desertification problem. If it is, we have a real problem. Now whether we want to spend $450 billion dealing with that is another matter. It does seem to me that we must try to get the answer to what is that relationship there because after all we do have a major change in that ecological system.

Ellis: We have adequate evidence to suggest that pastoralists and livestock-raising people are not responsible for an advance of the Sahara. My colleagues are reluctant to say that it's not advancing, and it's going to take 40 years of data to be able to say that definitely. Since it's their work, I feel less constrained in drawing conclusions. Their work demonstrates that the evidence that was presented in the 1970s—that the desert is advancing at 6 kilometers per year—is simply not true. So I don't think we have to wait 40 years to draw that conclusion.

Also I would point out that in much of Africa, wildlife resources and biodiversity are in fairly good shape. Some of the places in which that is most notable are where livestock-raising people live. Therefore, I think that livestock development and the maintenance and even improvement of biodiversity are compatible objectives.

Busby: A colleague of mine at the National Research Council has a concept called "ratcheting down of ecosystems." He says that as you go through a cycle of land deterioration caused by overuse, drought, or other factors, the productivity and regenerative capacity of the land does not return to its previous level, and you are a little worse off after each cycle. If he is correct, ratcheting down of ecosystems over time is something we should be very concerned about. The question is whether the Sahara will expand at 7 kilometers per year during the next drought and extend twice as far.

Batie: My Virginia Tech colleague, Cornelia Flora, constantly reminds me about the importance of smaller animals in the livestock system. Specifically she uses guinea pigs in Latin America as an example of the importance of assessing the reaction of women to suggested "improvements" in the cropping system that may imperil the use of weeds as fodder for the guinea pigs. Guinea pigs are a cash crop and provide status for women. We didn't hear anything about these kinds of interactions: either the role of women or the role of these smaller animals in the system. Would you please elaborate?

Ellis: If you're including small ruminants, I can answer. One of the analyses that we did in Turkana was to look at the role of the six species of livestock those folks raised. If you follow the way the energy flows through the different species of animals, you could see that each occupies a different and important functional niche. For example, cattle are only important during a short period of the year because that's the only time they give a reasonable amount of milk. On the other hand, that's the only time of the year when people gain weight, when women get pregnant, and things of that nature. So, the niche is a small one for cattle in that system, but it is an extremely important one. Camels on the other hand give milk year-round. So they're sort of the livestock that the people really depend on. But it looks as if human reproduction and demography depend very much on a bulge of energy that comes for a short period of time from cattle. You can go through each of the five species and describe their niche the same way. So diversity of species is very important.

The people that I have worked with don't raise guinea pigs or anything smaller than goats or sheep so I can't really respond to that issue. On the question of gender, in my experience in livestock systems, the women do about 90 percent of the work and the men about 10 percent, and the men make about 90 percent of the decisions with respect to livestock allocation. However, actual dispersion of food within the family is a female prerogative most of the time.

Maner: In the systems that we've looked at in northern Mexico, animals smaller than the goat or sheep are not very prevalent. In Latin America, most of the guinea pigs are still concentrated in the Indian communities in the high altitudes of the Andes in Peru, Bolivia, and southern Colombia, where the production of the guinea pig has been very popular and has been basically controlled by women. Women do most of the collection of forage. They do most of the production of the guinea pig within the house or near the house.

In northern Mexico, women play an important role in the pastoral system of the ejido. Men normally graze the goats during the day, coming in at noon and going back out in the afternoon. Women are responsible for the cheese-making and the processing of the milk more than anything else. They do not play a significant role in pastoral management within the ejido system.

Nyaribo-Roberts: In the high-potential agro-pastoral areas of Rwanda-Burundi, there's a high male out-migration due to small land holdings, and most of the production falls in the domain of women. My belief is that 15 to 20 years from now women will be playing a greater role in the care of large stock that are currently identified with men.

Kellogg: The CGIAR has done some economic analysis that has been conducted on equity issues. If you take a look at livestock products, you understand numbers of producers, numbers of consumers. Then you look at food, small grains, foodgrains, number of producers, number of consumers, and you understand the elasticity of demand for these two different kinds of products. Using the magic of economics, you can show that more equity is achieved—poor people are served better by increasing the supply of foodgrains than by increasing the supply of livestock. That kind of argument is one of the reasons why people have concentrated a lot more on crops than on agricultural livestock.

A question for Jim Ellis: You implied that there was more knowledge about how these systems worked and maybe about how one could work with them to improve the livelihoods of the people, but that development practice hasn't caught up because for many development agencies, livestock projects are just not in the cards and are not being implemented. Even projects in drylands and semi-arid areas are not popular. What things might be pursued that could make our development activities more effective in that part of the world.

Ellis: First we have to do is get over the idea that these are innately destructive systems. Second, ask: What's good about these systems? What works and what doesn't seem to work? Taking account of the successful indigenous aspects of this system, how can we, as outsiders, help improve the welfare and lifestyle of these people without debasing the successful aspects of the system as they exist today?

That process is under way at the moment. There have been a series of workshops sponsored by the Commonwealth Secretariat and Overseas Development Institute in London that are asking, how can we take our new knowledge and apply it to better development and management activities for range and livestock systems in Africa? I can tell you that the answer is not yet written down. There's another workshop coming up. It seems to me that we're focusing mainly on institutional responses, such as market systems and other sorts of institutions. In other words, it's not a question so much of educating the people themselves because we figure they know what they're doing. It's a question of developing institutions that are responsive to (1) the inherent dynamics of the systems and (2) the indigenous knowledge and what's gone right with these systems. The current national institutions do not do that. They don't take account of either the inherent dynamics of the system or what the people themselves, the livestock-raisers, are doing. How can we redesign those national and international institutions to make them more effective and responsive?

7
Genetic Conservation and Biodiversity

Introduction

ROBERT HORSCH
Monsanto Company

I have been with Monsanto for 12 years and my expertise is in the biotechnology domain of genetic engineering proper—transfer of genes. This is my view of the biodiversity situation: We're just at the beginning of being able to use biodiversity in a technological sense. This has been connected in the public's mind because of the debate over the Biodiversity Treaty at the Earth Summit in Rio de Janeiro last year and the unwillingness of the United States to sign it because of the implications for the biotechnology industry. There certainly are economic and technical linkages but, in some sense, it's premature to make that the justification for worrying about biodiversity.

Looking at some recent headlines, I was surprised how little we know about what's out there in terms of different forms of life. These were all headlines in the last few weeks:

- The world's first poisonous bird was discovered—the bird was known for a long time, but was not known to be poisonous.
- The world's biggest bacterium was discovered, big enough to see with the naked eye—this was not known before last month.
- Hope for an AIDS cure rose with an African vine—this is really the economic linkage of biotechnology and biodiversity: the search for pharmaceuticals and other compounds from natural sources.

The hypothesis I'll put forward is that habitat preservation is the biggest single key to the preservation of biodiversity. Agricultural productivity is certainly a major factor in reducing the pressure on habitats. The struggle to preserve genes and species by putting them into banks is certainly a worthy endeavor, but I would liken it to taking photographs of the world's great art and then letting the art be destroyed. The main access that the public would have then would be that they could look at the pictures, but probably they would look at a list of what was in the picture books. They would say, "woman smiling," and that would be their relationship to the Mona Lisa. I think having the organisms out there doing what they normally do is really what we want.

Genetic engineering is just one of the many keys, and it is unproven. It's a potentially a way to keep the advancement of agricultural productivity rolling. Just to give an idea of what makes up that promise: The list of crop species that can be genetically engineered has grown daily since the first tobacco and petunia plants in 1983 to the success with wheat in 1992. Field testing is an indication that the stuff is getting close to reality. Most of the early tests were with pest control for insects or viruses. They were followed by a large number for herbicide resistance, which is more a reflection of what is technically feasible than where the big economic opportunities are. Now there are tests on a growing array of food quality projects and the potential for producing molecules like bioplastics.

There have been more than 400 field tests done around the world. Most of them have been done by private companies, and most of those have been done by Monsanto. This certainly does not indicate a monopoly on the market. It just indicates an early lead.

The last point I want to make is that the patent system is what allowed Monsanto and other companies to make the decision to invest in this technology. It is what allows me and my colleagues to publish what we work on. And it is what allows other institutions that couldn't afford to get over the hurdle of cracking the technology in the first place to learn from what we've done and to apply it at a much faster rate and much lower expense in hundreds of diverse research institutions around the world.

This is a high-risk game. Many companies are no longer with us in the field of genetic engineering. There has been a lot of initial investment, a lot of venture capital has gone down the drain, but a lot of results and a lot of patent applications have been left behind.

At Monsanto our strategy for helping developing countries has been to ask, what are the targets? and then build the technology transfer around specific traits and targets and do it locally. In other words, our attempts at technology transfer have been to work with scientists from the developing countries and to involve the institutions in the developing countries, rather than trying to do the projects for them.

Genetic Conservation and Biodiversity: Key Issues for the Future

CLIVE JAMES
International Service for the Acquisition of Agri-Biotech Applications

AVTAR KAUL
Winrock International

Earth is the only celestial body where life is known to exist and, to our knowledge, the only habitable planet where humans have lived, survived, and evolved. Living organisms will continue to survive on this planet only if we can sustain its unique life-support system. Until recently, human beings were a relatively small entity in what appeared to be a very large home called earth. But we are now becoming acutely aware that this is no longer the case. Our fast-growing global family is outstripping the capacity of our global home. Mother earth, which we may have mistakenly taken for an infinite life-support system, is, mainly because of us, running out of time, and its biological diversity is being depleted through inadequate genetic conservation. This irreversible action threatens the ability of future generations to continue to exist, evolve, and enjoy an adequate quality of life on this planet (Dooge et al. 1992).

Gro Harlem Bruntland, the prime minister of Norway, has noted that throughout history the responsibility for the future of our children and grandchildren has always been part of human nature (Bruntland 1992). It has always been the pride of the current generation to pass on a heritage, a legacy, and a foundation on which the next generation can build. Many believe today that if the loss of biodiversity continues unchecked and if the global community is not prepared to take urgent concerted action to build new coalitions and partnerships for collective and equitable action, then we may be putting at risk the long-term possibility for life as we know it to continue to exist on this planet.

Acknowledgment. The authors gratefully acknowledge the assistance of Nyle Brady, Ronnie Coffman, Donald Duvick, Richard Flavell, Anatole Krattiger, M. S. Swaminathan, Calvin Qualset, Anne Westman, and Garrison Wilkes.

Conservation of biodiversity is increasingly being recognized as a requisite for sustainable development in such documents as *The World Conservation Strategy* (IUCN 1980), *Our Common Future* (World Commission on Environment and Development 1987), and *Caring for the Earth: A Strategy For Sustainable Living* (IUCN 1991). A common theme in all these analyses is the emphasis on the fundamental importance of the conservation of genes, species, and ecosystems to achieve sustainable patterns of resource management. They emphasize that biodiversity conservation covers the entire range of life forms including agricultural crop and forestry species, fisheries, livestock, and microorganisms. The goal of biodiversity conservation is to attain sustainable development by making use of and conserving biological resources in a manner that does not diminish the world's variety of genes and species and does not destroy habitats and ecosystems (Miller, Reid, and Barber 1992). A global biodiversity strategy has been developed through a series of consultations, workshops, and involvement of more than 500 individuals from around the world (World Resources Institute 1992).

While the process of development tends to emphasize only the tangible economic benefits, the general public recognizes the intangible values that enrich the quality of life, including social, ethical, and aesthetic factors and ecosystem services such as watershed protection, photosynthesis, regulation of climate, and protection of soil (Miller, Reid, and Barber 1992).

As a preface to discussing the topic of genetic conservation and biodiversity, it is useful to define biodiversity and also to briefly describe its significance. The U.S. Office of Technology Assessment (OTA 1987) has defined biodiversity as follows:

> Biological diversity, abbreviated to biodiversity, refers to the variety and variability among living organisms and the ecological complexes in which they occur. Diversity can be defined as the number of different items and their relative frequency. For biological diversity these items are organized at many levels, ranging from complete ecosystems to the chemical structures that are the molecular basis of heredity. Thus the term encompasses different ecosystems, species, genes, and their relative abundance.

The genesis of life began about 3.5 billion years ago, which is about 1 billion years after the origin of earth. The first terrestrial communities have probably existed for about 430 million years, and forests were established about 300 million years ago. These ecosystems have evolved and become more diverse over millions of years. Because of different growth rates, the tropics are the most species-rich area of the world, followed by the temperate regions. The Arctic zones are the species-poor areas (Arroyo, Raven, and Sarukhan 1992). It is estimated that 80 percent or more of the biodiversity in the world is to be found in the tropics, with about half in the tropical rainforests. Thus, the species on earth today have evolved biologically and culturally in a highly diverse world. The interactions between genetically diverse organisms has created evolution, and hence our future cannot be isolated from the genes, species, and the ecosystems that have and will continue to shape the form, fate, and future of life on this planet. That is why biodiversity merits our urgent attention.

Increasing Awareness of Biodiversity Issues

Three factors have probably been responsible for precipitating a new awareness in today's world about diversity.

Global population and growth rate

The global population is expected to double in the next 50 years, or so, to 10 to 11 billion. This doubling of the population will almost certainly result in a degrading of the environment at an unprecedented rate, and the loss of biodiversity will be especially important in tropical areas. Of today's human population of 5.5 billion, 77 percent are in the Third World, consuming only 15 percent of the world's resources and sharing only 15 percent of the global wealth. In contrast, the industrial countries with only 23 percent of the global population consume 85 percent of the world's resources and control 85 percent of the world's wealth (Arroyo, Raven, and Sarukhan 1992). This situation, considered inequitable by many, underpins one of the major controversies between the haves and the have-nots and impacts directly on biodiversity.

Loss of habitats and biodiversity

Much of the loss in biodiversity is through the extinction of natural habitats and most of this is happening in the tropics. For many, the denuding of the tropical forests, which contain approximately 40 percent of all biodiversity, has come to represent the loss of biodiversity on a global basis, and is of particular significance for the developing world. At their peak, tropical rainforests probably occupied about 12 percent of the globe. Today they occupy about half of that area, and it is estimated that between 1950 and 1990, the tropical rainforest area was reduced by 30 to 40 percent. Projections for depletion during the next 30 to 50 years suggest that another 30 to 40 percent of the forests will disappear, which could result in a 20 percent loss of global diversity, equivalent to 2 million species, or an average loss of 200 species per day (Raven 1987, Wilson 1989, Ehrlich and Wilson 1991, BOSTID 1992). A more alarming projection by the World Resources Institute (1990) predicts that if the deforestation rates of the 1980s continue, all tropical moist forests will be lost in less than 40 years. Examples of locations where whole habitats have been lost are the Narmada Valley dam in Gujarat, India; Nam Choan dam in Thailand, and the Grande Carajas program in Amazonia, Brazil. The discrepancy between estimates reflects the difficulty in projecting from an inadequate database and forecasting an uncertain future. However, the most important dimension of the extinction of habitats and biodiversity is that, whenever and wherever it happens, the loss is irreversible.

The interaction between biodiversity and biotechnology

The advent of biotechnology has opened opportunities for science and technology to better capitalize on the power of biodiversity, which in turn can be used to alleviate human suffering in the Third World and to ameliorate environmental degradation. Paradoxically, the nexus of biodiversity and biotechnology has also exposed another dimension of the controversy between the North and the South relating to ownership of biodiversity. The issues of intellectual property rights and the sharing and transfer of biotechnology from developed countries to developing countries is discussed later.

Principal Causes of Biodiversity Loss

The six most important causes of biodiversity decline were described by the World Resources Institute (World Resources Institute 1992a).

Unsustainable global population growth and consumption

World population is expected to continue to increase for at least another 50 years, with approximately a billion people per decade being added during each of the next three decades. Today people consume or destroy approximately 40 percent of all photosynthetic productivity, which is the source of energy for all living forms. There is a limit to the earth's ecological capacity to support a growing human population that increasingly diminishes biodiversity by destroying habitats, generating pollutants, depleting the ozone layer, producing acid rain, over-using nonrenewable resources, and wasting energy.

Economic undervaluation of biodiversity

There are several reasons for the economic undervaluation of biodiversity resources, particularly in developing countries. Many biological resources are consumed directly and are not accounted for in economic valuation. For example, in tropical forests, timber and pulpwood are valued at market prices whereas wood gathered by peasants for fuel and plants gathered for medicines are assigned no value. Biodiversity is, by and large, "public goods" resulting in undervaluation that governments use to justify offering tax incentives to increase the value of the habitat for other "more valuable" uses; this in turn can be detrimental to biodiversity. For example, tax incentives to stimulate conversion of wetlands to farm land for greater market value often do not yield the anticipated gains in value, and furthermore a substantial investment in civil engineering works is often required to reestablish the coastal protection formerly provided by the wetlands. Uncertainty over property rights can be a disincentive that encourages exploitation rather than protection. For example, a farmer will not plant a woodlot if he lacks property rights and is uncertain who will own the trees in 5 or 10 years. Similarly, forest dwellers who do not receive adequate protection for informal property rights, as compared with formal property rights for land in an urban area, tend to exploit rather than protect forest reserves. Generally, economic under-valuation leads to the perception that conservation represents a cost, not a potential profit, resulting in little protection for biodiversity resources.

Inequity in ownership

The ownership or control of most of the land by a minority of the population, as occurs in most of the Third World, promotes short-term profits as opposed to long-term sustainability. The situation is accentuated if the powerful and wealthy minority live in urban areas with little or no contact or sensitivity to the depletion of biodiversity in the rural areas. There is also inequity of ownership between the rich and the poor countries in the context of international trade, debt, and technology. This inequity has resulted in a reversal in the net transfer of funds from industrial countries to the Third World. In 1980 the flow was $42.6 billion in favor of the Third World, but by 1988 the flow had reversed to $32.6 billion in favor of the industrial countries. If the Third World continues to be denied technology, deprived of markets, and burdened with debt, it will neither have the means nor the incentive to conserve biodiversity for the future.

Narrowing spectrum of traded natural products and deployment of improved germplasm in monocultures

The current global exchange economy, which is based on comparative advantage and product specialization, has increased both the uniformity of products traded and interdependence between the industrial countries and the

Third World. In agriculture, producers tend to limit production to those few commodities that have comparative advantage in the world economy, and this has the indirect effect of depleting biodiversity within crops in cultivated areas. This effect is illustrated by the dramatically increased yields from improved semi-dwarf wheat and rice varieties, which have provided food for the millions, particularly in Asia, that otherwise would have gone unfed, but they have also contributed to narrowing of the germplasm base. Since 1980, 4 out of every 10 Asian farmers have adopted improved varieties of rice and wheat. In India 75 percent of the rice area is believed to be dominated by about 10 modern varieties; in Indonesia it is estimated that 1,500 local varieties may have been displaced by improved rice varieties in the last 15 years. The expansion of monoculture is of critical concern in forestry too. It is claimed that the Tropical Forestry Action Plan has contributed to the rapid spread of fast-growing eucalyptus monoculture in much of Asia, Africa, and Latin America, leading to erosion of native species. When extensive areas are under monocultures of tree or crops, the narrow germplasm base can increase the vulnerability to pests and diseases. This was the cause of the *Helminthosporium maydis* epidemic in maize in the United States in 1970. On the other hand, history records that stem rust periodically devastated traditional wheat production in different parts of the world since biblical times, but since the 1960s, 50 million hectares of semi-dwarf wheat, representing 70 percent of all wheat grown in the developing world, excluding China (CIMMYT 1992), has not suffered a single epidemic of this devastating disease.

Institutional and legal constraints

Lack of government institutional capacity and an unwillingness and inability to formulate cross-sectoral programs are major constraints to halting the depletion of biodiversity. Consequently the implementation of programs is fragmented. Scientific disciplines operate in isolation and project-specific activities are not managed within the context of the boundaries of a "biodiversity region." But conservation of ecosystems is often dependent on critical factors external to the specific habitat and should therefore be an integral part of regional development planning. In many developing countries, the institutional constraints are exacerbated by a lack of environmental laws to protect diversity. Even where laws have been enacted, they often cannot be enforced. Inadequate institutional capacity is possibly the most important barrier to the implementation of effective biodiversity programs on a global basis.

Deficiency in information and knowledge

International knowledge of the status of biodiversity, the rate of depletion, and the functioning of the various ecosystems and their innumerable components is poor. An extensive research effort will have to be conducted simultaneously with project formulation because of the need to act quickly to implement projects to try to halt the extinction of habitats. However, lack of knowledge is seriously affecting the efficiency of decision making, project formulation, and long-term planning, which has to be continuously revised as new knowledge dictates different solutions and approaches. Lack of knowledge can also undermine the credibility of the case for diversity in the view of the political, professional, and lay public communities.

Development and Biodiversity

Development agencies must take a leading role in the debate over genetic conservation and biodiversity for four major reasons. First, 80 percent or more of global biodiversity is found in the Third World, which is the focal point of development agencies. Second, development agencies implement projects that can have a direct impact on biodiversity. These include dam construction, forest clearing, road construction, transmigration and resettlement projects, and the large-scale introduction of improved crops such as rice and wheat. Whereas there is no doubt that such projects have led to the narrowing of biodiversity, monitoring the impact on biodiversity in projects will allow development agencies to benefit from this and other knowledge to design projects that do not deplete biodiversity and that, it is hoped, will maintain biodiversity in future (Kaul 1989). Third, because the economies of most Third World countries are likely to remain weak for a long time, the international development community is likely to be the only major source of funds for biodiversity activities in the Third World. Finally, the international development community represents an interface between developed and developing countries and given that the prevailing view is that conservation of biodiversity is a global responsibility, development agencies may be the most appropriate institutions to play the role of honest broker.

Biological Considerations

One of the difficulties in dealing with genetic conservation and biodiversity is that we know so little about biodiversity. Only about 10 percent of all species have been classified, so the database for decision making is hardly adequate, making the formulation of development strategies to address biodiversity extremely difficult. Past projects to conserve biodiversity have demonstrated that integrating scientific, economic, social, and political action, based on a sound knowledge of genes, species, and ecosystems, is essential for success. In reality, however, much of this information and knowledge is lacking, and at the same time there is a great urgency to act because of the accelerating rate of loss of biodiversity, which is irreversible (BOSTID 1992).

The inadequacy of biological surveys, inventories, and monitoring programs to characterize the different dimensions of biodiversity and to generate knowledge precludes reliable quantification of the status of biodiversity, the assessment of needs, and the assignment of priorities. The best estimates suggest that 10 million species may exist. Approximately 1.4 million have been described, of which 0.75 million are insects, 0.25 million are plants, and 0.44 million are vertebrate animals. It is estimated that given current rates of loss of biodiversity, 20 to 25 percent of the world's plant species will be lost and more than a third of those in the tropics may be at risk during the next three decades. Only about one-fifth of the total area of global freshwater habitats has been documented and less than 1 percent of marine species and deep-sea habitats have received due attention (Arroyo, Raven, and Sarukhan 1992).

In summarizing current knowledge about global biodiversity, E. O. Wilson concluded that "we cannot confidently estimate the number of species of organisms on earth even to an order of magnitude," which he views as an appalling situation (Wilson 1988). Arroyo, Raven, and Sarukhan (1992) further note "that there are clearly few areas of science about which so little is known, and none of such relevance to human beings." In the face of our ignorance of the basic facts, one evident challenge for scientists will be to define a theoretical context for biodiversity in order to address emerging issues in this area and to develop realistic sampling procedures to collect badly needed additional data

Conservation is the key action for saving biodiversity. Genetic conservation can be *ex situ* or *in situ*. Ex-situ genetic conservation is achieved through storage of seed or tissue in germplasm banks, and, in the future, it will be possible to store genes in a gene bank. Although there is a network of national and international seed banks for important food, feed, and fiber crops, knowledgeable observers such as Wilkes (1992) consider the extent and the efficiency of these ex-situ germplasm banks to be inadequate. The ownership of this germplasm, most of which has come from the Third World, is a subject of continuing controversy between the developed and developing countries and continues to be debated in forums such as the FAO Commission for Genetic Resources. Political actions and financial stringencies that jeopardize the safety of the genetic resources stored in formal and informal germplasm banks may threaten biodiversity as much as environmental degradation does. Examples are Sudan, Afghanistan, and Somalia, countries wracked by famine and civil war, where formal germplasm banks have been deserted and looted while starving farmers have been forced to eat their seeds. These countries have virtually lost the indigenous seed resources that for centuries farmers had selected for adaptation to the local conditions. Fortunately for Somalia, just before it fell into anarchy, an IBPGR staff member hand-carried 300 samples of sorghum and maize to Kenya. These along with some other exotic samples will be repatriated to rebuild the Somalia's seed base (M. S. Mengeshe, personal communication, 1993).

Few in-situ protected areas have been established to conserve genetic resources and biodiversity in their natural ecosystems. Moreover, there are few, if any, protected ecosystems modified by human beings. This latter point is important because the challenge is to protect biodiversity on agricultural land where there should be a healthy relationship between biodiversity and sustainable production.

The Keystone conference on global initiatives for the security and sustainable use of plant genetic resources (Keystone Center 1991) suggested strengthening of efforts in the following activities:

- Ex-situ conservation: collection, storage, regeneration, documentation, information, evaluation, exchange
- On-farm community conservation and utilization
- In-situ conservation
- Monitoring and early warning of genetic erosion in specific locations
- Development of techniques for sustainable advances in agricultural productivity
- Research, training, and public education

Another challenging biological issue is the restoration of ecosystems that have become degraded, but the knowledge and technology to achieve this is incomplete and has yet to be put into practice.

Plant and microbial biodiversity has been demonstrated to be important to human health in both traditional and modern medical practices. Plant, animal, and microbial varieties are crucial to the healing process of traditional medicine, which forms the basis for primary health care for about 80 percent of the people in the developing world (Farnsworth 1988). More than 5,100 species are used in Chinese traditional medicine alone, and the rural communities of Iran, Pakistan, Afghanistan, India, Nepal, Bangladesh, Myanmar, Thailand, Vietnam, Cambodia, Indonesia, and Philippines depend on herbal medicines as much as the Chinese. Over 25 percent of the prescriptions dispensed in the USA contain active ingredients from plants, and 3,000 antibiotics are derived from microorganisms. Compounds extracted from plants, microbes, and animals are involved in developing all the 20 best-selling drugs in the USA. The combined sales of these drugs approached $6 billion in 1988 (Miller and Brewer 1991).

The Dilemma of Formulating Effective Biodiversity Projects

Biodiversity represents a dilemma for developing countries for several reasons. First, it is difficult for a country to focus and devote resources to long-term needs when it is faced with critical immediate needs for food, feed, fiber, fuel wood, and some way to generate foreign exchange to buy essential products and repay foreign debts. Second, the countries of the Third World are the poorest nations on earth, with large and rapidly growing populations, rudimentary infrastructure, and inadequately trained human resources, but they possess approximately 80 percent of the world's biodiversity, and they do not have the capacity to conserve it effectively. Yet without prompt initiation of steps to halt the depletion of biodiversity, the total global community will lose the genetic resources that will be essential for sustained economic development. Conversely, better conservation of genetic resources and protection of biodiversity is not likely to take place in the developing world unless there are quantum leaps in economic development.

A USAID report on a project in Thailand illustrates the problem of dealing with biodiversity for both the developing country and the aid agency (U.S. Agency for International Development 1986):

> In Thailand, the principal threat to long-term maintenance of biological diversity and tropical forest resources is agricultural encroachment on already-designated conservation areas by nomadic hill tribe groups and landless lowland Thais. Overcoming this problem is fundamental to the long-term viability of much of Thailand's biological resources, and will require concerted efforts in agricultural, rural, and economic development, as well as in reforestation and protected area management. For those situations where basic human needs must be met for conservation efforts to succeed, USAID might be able to play a role both in conserving biological diversity and tropical forests and in fostering sustainable economic and social progress of the poor.

Brady (1988) noted that in order to develop a solution, ways would have to be found simultaneously to

- conserve more natural habitats
- better manage those that already exist

- ensure that development projects are ecologically sound
- employ improved methods of economic analysis of the costs and benefits of natural resources deterioration and investments in natural resources conservation
- increase food and fuel wood production on land already cleared to reduce pressure on the remaining wild areas.

That is quite a challenge.

The International Biodiversity Convention

Access to and ownership of biodiversity are the major issues related to the primary genetic resources in the Third World. Similarly, access to and sharing of profits with developing countries from the higher valued biotechnology products of the industrial countries continue to be the most significant stumbling blocks in the dialogue between the North and the South on biodiversity, However, encouraging progress was made at the United Nations Conference on Environment and Development in 1992, when the Convention on Biological Diversity was signed by more than 150 countries, including all industrial countries with the important exception of the USA. Two issues stimulated the countries to sign the treaty. The first was a recognition by the overwhelming majority of nations attending the conference of the need to conserve habitats and accept biodiversity as a global responsibility. The second was the acknowledgment that biotechnology was conferring added value to genetic resources, resulting in a financial value expressed through intellectual property rights. The treaty must be ratified by 30 or more of the 150 signatory countries before it comes into effect. Canada was the first major industrial country to ratify the treaty, and it has been joined by about 10 developing and other smaller industrial countries. The likelihood is high that the treaty will now come into effect in the near term, particularly given the announcement by President Clinton on April 20, 1993, immediately prior to Earth Day, that the USA has reconsidered its position and now plans to sign the convention. Recognizing that the signing of the treaty is only a first step, that its language is ambiguous, and that it does require important implementing legislation to have an effect at the grassroots level, it is nevertheless an important international treaty that addresses biodiversity. Moreover, it addresses the conservation issues that deal with access, technology transfer, and intellectual property rights and, therefore, a closer examination of the treaty is worthwhile. The intent of the treaty (IUCN 1989) is:

> to serve as a key coordinating, catalyzing, and monitoring mechanism for international biodiversity conservation. It will also be the primary means of establishing accepted international norms for biodiversity conservation. Although current international agree-

ments cover some elements of biodiversity conservation, taken together they do not cover all the world's threatened biodiversity, and they do not adequately address the closely related issues of use, ownership, funding, and technology transfer.

The foundation of the convention is a commitment by members to conserve biodiversity, but this will probably be a delicate clause to implement because it will require a developing country to surrender a national sovereignty right over biodiversity in exchange for a financial commitment from a donor to protect a designated ecosystem or species. Prior to the Rio conference, the draft treaty had specific lists of threatened species, but these lists, which provide teeth for the preservation of conservation, was contested and removed before the final draft was signed in Rio.

The article in the convention associated with the economics of genetic resources was the primary basis for the United States' rejection of the convention because it was believed that it would be detrimental to negotiations for stronger protection of intellectual property rights at the Uruguay Round of GATT.

Although the treaty language is general and in some cases ambiguous, the convention does represent an important step forward and indicates an intent by the parties and is a point of departure for further and more specific dialogue. The language of the treaty is clearest in relation to access to genetic resources when it grants the sovereign rights to nations over their genetic resources, with acquisition details (costs, technology transfer, etc.) to be agreed upon by the parties on a case-by-case basis. Thus, an important change, following the ratification of the convention, is that free or scientific accesses to genetic resources, which previously played an important role, may be terminated. Central American countries and the Southern African Development Coordinating Conference are now drafting legislation to implement this article of the convention. Similarly, India is formulating policies for the conservation of biodiversity (Swaminathan Research Foundation 1993). However, genetic resources already obtained from a nation (for example, germplasm already in storage in a national or international germplasm bank) is *not* included in the treaty and can continue to be used freely as before.

The article governing intellectual property rights is more ambiguous but nevertheless it does represent progress. Barton (1992) has summarized the outcome of the treaty in relation to intellectual property rights as follows:

> The developed nations have agreed to provide technologies relevant to conserving and using genetic resources "on fair and most favorable terms, including on concessional and preferential terms where mutually agreed." These terms however are to be consistent with intellectual property legislation but the parties will cooperate "to ensure that such rights are supportive of and do not run counter to [the convention] objectives."

There is a special article (no. 19) in the convention that deals with biotechnology that requires, as appropriate, that countries that provide genetic resources be given the opportunity to participate in biotechnology research. It also encourages access to the results and benefits of biotechnologies based on genetic resources provided by the parties (Barton 1992).

Significant progress also has been made on financial commitments in that developed nation signatories committed "to meet the agreed full incremental costs to [developing nations] of implementing measures which fulfill the obligations of the convention" (Barton 1992). The negotiations on finances centered on two issues: the amount of funds to be committed and the degree of control on funds by donor countries. The general agreement was to reinforce the long-standing commitment to pledge 0.7 percent of GDP as aid, but a much more important subsequent interim measure decision was to fund biodiversity activities through the donor-controlled Global Environment Facility. The GEF is a $15 billion fund set up for a 3-year pilot program addressing four priorities: loss of biodiversity, global warming, pollution of internal waters, and destruction of the ozone layer. The fund is managed by the World Bank with the participation of UNDP and UNEP. It recently set up panels to study four aspects of biodiversity—conservation, economics, technology transfer, and biosafety. Whereas coordination and follow-up of all the conventions, including biodiversity, falls under the purview of the newly established United Nations Sustainable Development Commission, funding of environmental activities including biodiversity will come under the control of GEF.

The major challenge for the GEF, headed by Mohamed el-Ashry, is in his own words, how to play "a catalytic role by integrating global environment considerations into the regular development assistance programs of bilateral and multilateral donors" (Expert panels 1992). GEF must deliver resources to fund well-conceived projects that generate new knowledge and results at the grassroots level in a reasonable time frame and with the minimum of bureaucracy. In other words, action is infinitely more credible than words particularly when the need is as urgent as it is with biodiversity.

Role of the CGIAR Centers

The work of the Consultative Group on International Agriculture Research (CGIAR)[1] is based on the recognition that genetic resources are fundamental to the development of improved, more productive, stable, and sustainable sys-

---

[1] Summarized from a unpublished policy paper jointly prepared by the CGIAR centers, April 1993.

tems in agriculture, forestry, and fisheries. Thirteen centers are directly involved in the conservation and exploitation of plant genetic resources, three are involved with livestock, and one with fish genetic resources.

Though the CGIAR was not eligible to sign the biodiversity convention, it will significantly affect the functioning of CGIAR centers. The centers collaborate with a wide range of institutions in the public and private sectors in countries that vary in size, strength, capability, and above all in the extent of their natural endowment with genetic resources as well as the extent of exploitation of plant genetic resources and security of conservation. These differences must be clearly recognized in the further development and implementation of any global systems concerned with the conservation, sustainable use, and equitable sharing of the benefits of exploiting biological diversity. The issues go beyond the view that developing countries hold the majority of genetic resources while developed countries hold overwhelmingly the means to exploit these resources. The CGIAR philosophy is that it is inappropriate to treat species on which humanity depends for food and other basic necessities (CGIAR mandate crop species) in the same way as species of high industrial value, such as those of pharmaceutical potential. The challenges of conservation and equitable exploitation of land races and farmer varieties of traditional food crops are vastly different than those for wild species that are of interest to the chemical industry.

In response to article 8 of the biodiversity convention concerning in-situ conservation, CGIAR centers can play an important role in strengthening national efforts through research, training, assistance in planning, and laying emphasis on the conservation of wild relatives of crop plants. In-situ and ex-situ conservation efforts can be integrated in a mutually complementary initiative. CGIAR centers recognize the importance of on-farm conservation of plant genetic resources and the role that farmers play in crop conservation and improvement. Contacts at farm and community level are considered valuable. Ex-situ conservation should be considered equally important and not as a backup measure to in-situ conservation. Collectively, CGIAR centers house approximately 509,000 accessions in their germplasm banks (Wilkes 1992). CGIAR centers consider themselves as guardians and *trustees*, not owners, of these collections. These genetic resources are held in trust for the world community, particularly the developing countries. Conservation, duplication, repatriation, and free distribution to end-users of germplasm are some of the services rendered by CGIAR centers as trustees. In addition, materials are well characterized and documented for use as genetic stocks. Therefore, the centers link conservation with use, an important objective of the convention. The International Board for Plant Genetic Resources (IBPGR), which does not hold any collections of its own, assists national programs through information dissemination, training, joint collection, and repatriation. The CGIAR offers opportunities to all end users to obtain germplasm without restrictions. Unrestricted access to diversity is the key guideline of CGIAR centers, and thus the CGIAR should function as a global system that provides complete access to germplasm.

Subsequent to the ratification of the convention, the CGIAR hopes to continue to receive the germplasm from all countries. In support of article 10 of the convention, which emphasizes sustainable use of biodiversity, the CGIAR centers uphold the concept of enhanced productivity through the genetic ability of crop plants to withstand pests, diseases, and adverse environments. The IBPGR in collaboration with other CGIAR centers and national institutes has trained almost 2,000 people from developing countries in aspects of plant genetic resources conservation. Therefore, the CGIAR strongly endorses article 12 of the convention.

The CGIAR endorses article 13 of the convention, which aims at increasing the understanding among decision makers of the role that plant genetic resources can play in national development. The centers believe that it is important that farmers' rights receive due recognition through a compensatory mechanism of an international fund. The breeders' rights are recognized through "limited restrictions." While CGIAR policy is that genes or genotypes from center collections are not patentable, it is currently involved in a collaborative study to look at ownership issues in greater depth. CGIAR considers it crucial that, at least for basic food crops, an absolute minimum of restrictions be applied in the exploitation of genetic diversity by the originators. The centers have already adopted a policy of not applying intellectual property protection measures to any of the materials they hold in trust. It is hoped that some conditions can be applied to recipients of the materials from their germplasm banks. The centers, in addition to distributing original plant materials, supply breeding lines originating from their own research along with technology and relevant information, including documentation and training materials. In support of article 18 of the convention, the CGIAR expects to make a significant contribution to national programs through scientific cooperation in the field of conservation and sustainable use of biological diversity.

The centers share new materials and new techniques involving biotechnologies with their national partners. Training of scientists from developing countries is an important component of this program. All national and international rules are followed in conducting biotechnology research. The CGIAR believes that it should be invited to be on the Subsidiary Body on Scientific, Technical, and Technological Advice proposed by the convention through article 25. It has valuable expertise to offer.

While the CGIAR subscribes to the objectives addressed at the Convention on Biological Diversity, the CGIAR recognizes that certain areas, particularly those that impinge directly on its work, require further discussions and negotiations. Two major concerns are access to genetic resources and technologies and the role of international ex-situ collections.

Impact of Biotechnology on Biodiversity

The advent of biotechnology in agricultural research during the last decade had important implications for developing countries and the international agricultural research centers that seek to assist them. In the past most genetic research was done by scientists in public institutions and they made genetic materials freely available to other scientists throughout the world. For example, the dwarfing genes, which were the basis of the Green Revolution in rice and wheat, were made freely available to IRRI and CIMMYT and were incorporated in adapted germplasm and then shared with developing countries. Access to genetic resources from the Third World was also easily and freely achieved through international and bilateral germplasm exchange programs.

However, unlike traditional nonproprietary agricultural technology, where the public sector was the major stakeholder, the private sector is the major investor in biotechnology, accounting for an estimated $8 billion of the $12 billion global annual expenditures on biotechnology research and development. The primary consequence of private-sector involvement, which incidentally has been very much encouraged by most OECD member governments, is that the biotechnology products produced by industry will be proprietary, and intellectual property rights will apply. They will also be probably higher value-added products because of the higher costs of research and development in biotechnology. This does not necessarily translate to higher costs of production because savings, for example, through less use of pesticides, also can be realized.

The immediate market for the proprietary biotechnology products of the private sector is the industrial countries. However, many if not most of these products do have important potential applications in developing countries. For example, many of the biotechnology applications that have already been developed, such as nonconventional resistance to virus diseases or insect pests of crops, are already being tested in the Third World. Deployment of a biotechnology application featuring a Bt gene that has the potential to control many of the major pests of cotton probably will be feasible soon. Cotton is very susceptible to insects so it requires more insecticide than any other single crop, and two-thirds of world's cotton plantings are grown in developing countries. This new technology could therefore make an important contribution to sustainability by providing a substitute for some insecticide applications. Indeed it may be one of the first examples that can demonstrate that the philosophy of sustainability can be translated into an actual technology that can be adopted in the Third World.

Many of the superior agricultural biotechnology products of the future will likely be proprietary, whether developed by the private or public sector. The challenge is to develop innovative institutional mechanisms and create joint venture opportunities between the developed and developing countries that will give the latter access to the new proprietary products of biotechnology.

The decision of the developing countries to accept, or not to accept, products of biotechnology has both political and policy implications, and should be taken within the broader context of the long-term consequences of applying biotechnology techniques. Long-term implications include opportunities to improve the quality of tropical crops so that the Third World can continue to compete effectively in export markets. Conversely, industrial countries may use biotechnology to modify crops, which in turn may substitute for crop products traditionally imported from developing countries. There is little doubt that in the long term such substitutes will become possible with biotechnology. Hence prudent decision making by developing countries in the short term should anticipate such long-term developments. Failure to make the correct strategic choices at the initial stage could lead to a greater disparity between industrialized and developing countries in terms of access to biotechnology products and applications.

The confluence of biotechnology and biodiversity brings together the conflicting concerns and interests of the developed countries now represented by both the private and public sector and the developing countries represented by national governments. The developing countries are acutely concerned about over-reliance or dependence on external biotechnology, particularly for strategic areas like food security. There is a concern that over-reliance on external biotechnology may also stifle the opportunity to build a sufficient domestic capacity in biotechnology. Continuing and possibly high costs associated with royalties or license fees is another important issue. There is great concern in developing countries that local genetic resources would be modified and patented by the industrial countries, thereby denying the country of origin access to its own genetic resources, albeit modified, or alternatively that industrial countries will limit access to modified genetic resources by setting high purchase prices. On the other hand, the private or public sector in the industrial countries will be concerned about the willingness of the developing countries to adhere to patent rights, about their ability to recover and repatriate revenues, and about the possible leakage of patented technology to other markets. However, as a science, biotechnology offers important new tools for the global community to use in conserving biodiversity. For example, biodiversity can be conserved at the molecular level, and Baylor College in Texas has just established a DNA Banks Network. Biotechnology products also are being developed to protect plants from pests and diseases, and therefore offer a sustainable substitute for conventional pesticides.

Developing countries and the private sector in industrial countries can be considered adversaries on biotechnology issues, but they could instead be partners because each possess what the other requires. The developing country has raw genetic resources in the form of biodiversity, which it requires for biotechnology, but usually Third World nations cannot develop the products because of inadequate human, material, and financial resources. Conversely biodiversity in the form of genetic resources is the raw material for biotechnology and the biotechnology developers are anxious to make their products available to both industrial and developing countries if equitable arrangements for technology sharing can be devised.

The International Service for the Acquisition of Agri-Biotech Applications (James 1991), founded in 1991, has demonstrated that partnerships can be built between the private and public sectors of industrial countries and governments in the Third World that allow the sharing of proprietary agricultural biotechnology applications. ISAAA is a not-for-profit organization, and its institutional structure is unusual in that it is jointly sponsored by both public and private organizations, including philanthropic foundations, bilateral agencies, and private corporations.

Three ISAAA projects have been fully brokered, funded, and implemented, demonstrating the feasibility of transferring proprietary agricultural biotechnology to the Third World. The first model project involves the donation of coat-protein genes by Monsanto Company to Mexico for the control of viruses in the Alpha variety of potato. The project is funded by the Rockefeller Foundation and features technology transfer and the training of Mexican scientists. The first generation of transgenic potatoes, developed by Mexican scientists in less than 18 months, has now been field tested. A companion project is helping Mexico to develop the infrastructure and regulatory biosafety procedures for testing and introducing recombinant products. The second project, funded by USAID, involves the development of a cold DNA diagnostic probe by Washington State University for *Xanthomonas campestris* pv. *campestris*, which causes the most important disease of crucifers worldwide. The probe will be made available to the Asian Vegetable Research and Development Center, which in turn will make it available to countries in the Third World. The third project involves the development of nonconventional virus resistance (NCVR) to the cucumber mosaic virus of criollo melon in Costa Rica. This technology is being donated by the Asgrow Seed Company to the University of Costa Rica, and putative transgenics were generated in less than 6 months.

The ISAAA experience demonstrates that with trust and confidence new partnerships can be built to overcome the impasse that has existed between the developed and developing countries because of the controversies related to ownership of genetic resources and intellectual property rights. The ISAAA experience covers 10 target countries in Africa, Asia, and Latin America that have a political will to pursue biotechnology and have a measure of capacity in the science. To date, the projects are confined to donation of biotechnology, but there are obviously opportunities for building joint ventures between the North and the South involving a commercial interest for both parties. These are being explored.

The ISAAA experience in brokering the donation of proprietary technology between private corporations and the developing countries is not an isolated case. Merck and Company, the world's largest pharmaceutical company, concluded a $1 million partnership agreement with INBio of Costa Rica in 1991. Under the terms of the agreement, INBio shares with Merck plant and insect samples from its genetically diverse flora and fauna in return for payments and royalties, a portion of which are used to fund conservation of biodiversity in Costa Rica.

Several advanced developing countries that have already developed significant capacity in agricultural biotechnology are generating and testing their own transgenic products and are establishing regulatory oversight legislation to ensure that genetically modified organisms are tested in a responsible way in harmony with international standards. Furthermore some countries are contemplating various forms of patent protection that will be used to protect both imported technology and, perhaps more important, their own proprietary products in the future. These developments are encouraging; progress in the controversial area of ownership of genetic resources and intellectual property rights will ensure that developing countries will not be denied the opportunity to both access and generate the superior proprietary products that will become available in the international marketplace.

Institutional, Socioeconomic, and Cultural Issues

Institution building

There has been a quantum leap in global awareness about preservation of biodiversity. But institutional capacity throughout the world is inadequate to meet the enormous magnitude of the task. Education, scientific, and socioeconomic research institutions need to develop competence in this area to train people and generate knowledge. Public and private institutions need to develop an understanding of the issues related to all aspects of policy, particularly economic policy, so that meaningful planning can be undertaken. And the international development community needs to gather and utilize the information that will allow effective biodiversity projects to be formulated in the developing countries.

A failure to assign highest priority to strengthening institutional capacity in biodiversity in the developing countries will deny them the opportunity to build a stake in and commitment to biodiversity preservation that is a pre-requisite to long-term success in protecting biodiversity. Indeed institution building per se probably merits being assigned the top priority in the global biodiversity initiative.

Socioeconomic issues

A report of a panel of the Board on Science and Technology for International Development (BOSTID 1992) concludes that "the accelerated rate at which the world's biological diversity is being eroded can be attributed, in large part, to socioeconomic factors that encourage exploitive development practices while discouraging conservative resource use." Several observers of global biodiversity decline have called for urgent action by economists working in conjunction with agriculturalists, ecologists and development specialists to employ a more meaningful valuation of the real long-term economic loss associated with the irreversible extinction of an ecosystem or species. There is a majority view in the literature, articulated by Jeffrey McNeely and others, that biological resources are not valued by economists at appropriate prices and that "conventional measures of national income do not recognize the drawing down of the stock of natural capital, and instead consider the depletion of resources, i.e., the loss of wealth as net income" (McNeely 1988).

Some of the economic forces that impinge on biodiversity operate at the international level and others at the national and project level. Consequently the thrust of economic analysis on biodiversity should have a macroeconomic as well as a microeconomic perspective and should determine the nature of the interactions between the international and national level. Three critical areas have been identified for socioeconomic research (BOSTID 1992). First a careful study should be made of the causal mechanisms (taxes, royalties, land tenure systems) that often affect, or are an integral part of, development programs to determine their short-, medium- and long-term effects on the depletion of biodiversity. Second because the economic value of biodiversity is judged to be inappropriately assessed and undervalued as a resource, work should be undertaken to develop alternative valuations of biodiversity that reflect the long-term value of natural resources. Third, a study is needed on the disincentives that are leading to depletion of biodiversity and on options that would represent economic incentives to facilitate the replenishment of biodiversity.

Cultural considerations

If we accept that the depletion of biodiversity, by and large, results from social processes, it is reasonable to suggest that a study of the causal mechanisms and options for changing current practices might lead to the identification of solutions that could halt and, it is hoped, reverse the depletion. The complex social patterns in developing countries that are engaged in land clearing, slash-and-burn agriculture, logging, dam and road building, and transmigration need to be studied to gain an understanding of the incentives and disincentives that govern social thinking. The research done in this area is negligible, and the probability of formulating effective projects without this information and knowledge is small (Kaul 1990).

Another social aspect of biodiversity preservation that deserves urgent attention is that of the indigenous cultures that possess knowledge of the usefulness of species that have yet to be classified. Of particular importance are the medicinal uses of certain species by nomads and forest dwellers whose cultures are disappearing. It is noteworthy that in spite of scientific advances the active ingredients of most modern medicine are still based on natural products that have been derived from a minute fraction of global biodiversity resources. But medicine should not be the only focus of these research and information gathering activities. The work should include other practitioner groups such as subsistence agriculturalists for unconventional crops and animals, coastal fishermen, and artisans using natural resources in their crafts. Of particular importance are women in agriculture and their unique knowledge of biodiversity. In Africa women produce an estimated 70 percent of total food production, are intimately involved with the cultivation of crops, and are knowledgeable about their biodiversity and use in traditional cuisine (Rodda 1991, 180).

Global Biodiversity Strategy

There have been many proposals for future action. However the most comprehensive has been the Global Biodiversity Strategy prepared in 1992 jointly by the World Resources Institute, the World Conservation Union (formerly International Union for the Conservation of Nature), and UNEP in consultation with FAO and Unesco (World Resources Institute 1992a). The strategy, summarized in a policy maker's guide (World Resources Institute 1992b), is based on two premises:

- that successful action to conserve biodiversity must address the full range of causes of its current loss and embrace the opportunities that genes, species, and ecosystems provide for sustainable development
- that because the task is so broad, any biodiversity conservation strategy must also have a broad scope

Major thrusts

The strategy has three major thrusts—to save, to study, and to use biodiversity.

Saving biodiversity. The strategy concludes that the best way to save species is to protect their habitats. This can be achieved by preventing the degradation of natural ecosystems; by maintaining diversity of ecosystems modified by humans, which now represent the majority of habitats; and by restoring lost species to their former ecosystems including the preservation of species ex-situ in botanic gardens, zoos, and germplasm banks.

Studying biodiversity. Study will involve documentation of the composition, distribution, structure, and function of biodiversity at the gene, species, and ecosystem level. Building an awareness of the value of biodiversity and the opportunity for communities to participate in its protection through education will also be an important objective.

Using biodiversity. The strategy seeks to use biodiversity sustainably and equitably to demonstrate tangible benefits and involve a broad spectrum of individuals, indigenous peoples, women's groups, local communities, national programs, and international development agencies. Whereas a broad range of issues will be addressed including policy and law, human resources, and research, the emphasis will be on implementing an action program to protect threatened species.

The approach

The strategy defines five key objectives for achieving effective action.

1. Develop national and international policy frameworks to sustain biodiversity. The strategy concludes that current policies and practices are a disincentive to conservation of biodiversity, i.e., they cause economic under-valuation of biodiversity, and it will seek to reverse the policies in order to provide incentives. To achieve this objective, biotechnology will be used as a focus to demonstrate the increased value of genetic resources, and countries will be encouraged to acquire and foster biotechnology to build domestic capacity in the science. The Third World's US$1.47 trillion debt (International Monetary Fund 1993, table 5) will also be addressed as well as trade and other factors that currently hamper conservation and the sharing of biotechnology.

2. Provide incentives for effective conservation by communities. For sustainability, any successful biodiversity initiative must ultimately gain the support of the local communities who live with and are part of the biodiversity, which is largely found in the Third World. To do so, incentives and benefits must flow to the local communities that will actively play a fundamental role in the projects, and this could require complex issues to be addressed, including land tenure and adaptation of traditional practices, as well as major investments to be made in training for new skills.

3. Develop tools to conserve biodiversity. The strategy seeks to mobilize increased financial resources and personnel to strengthen ex-situ conservation but also to develop in-situ protected areas. A concerted effort will be made to implement biodiversity programs "bioregionally" to reflect ecological and social realities and to achieve a cross-sectoral approach and devolution of responsibility from central authorities to the local community level. The protected areas would constitute a national network and would be integrated with the bioregional programs, which would include natural and modified habitats. The latter is a new initiative that would designate conservation easements on private land. The strategy emphasizes mechanisms for increasing incentives and benefits to local communities through ecotourism, sustainable use of natural products, and protection of vulnerable species in off-site locations before their reintroduction to their former habitats.

4. Develop human capacity for conserving and using biodiversity sustainably. The strategy concludes that there is a chronic underinvestment in biodiversity training, particularly in developing countries. Investment in human capital is rationalized on the premise that conservation can succeed only if people understand, believe, and can manage biodiversity sustainably to meet human needs. The strategy envisions the establishment of national biodiversity institutes that would catalogue a nation's biodiversity, consolidate a database, and provide training opportunities for professionals for the multitude of tasks that the strategy outlines.

5. Achieve catalytic action through international cooperation and national planning. The current international cooperation effort is judged inadequate, and the strategy calls for an escalation of international cooperation to establish more demanding norms of conduct, increased mobilization of financial resources, and broader participation from the scientific and nongovernmental communities. The strategy acknowledges that international cooperation must be linked to national planning, which is considered the key that will allow biodiversity concerns to be incorporated into national economic development policy.

Catalysts for accelerating the strategy

The last step, to accomplish the goals of the strategy, focuses on five actions that in turn will act as catalysts for 80 actions in various sectors and institutions. The first of the catalytic actions has already been achieved with the adoption of the Convention on Biodiversity at the Earth Summit in Rio de Janeiro in 1992. The second is to seek United Nations designation of 1992-2003 as the International Biodiversity Decade during which the 80 action programs of the strategy would be implemented. The third is to establish an international panel on biodiversity, with broad representation, to develop international priorities for research and funding and a priority action list for endangered species, habitats, and ecosystems. The fourth catalytic action is to establish an early warning network to monitor threats to biodiversity and to alert the global community of the need for urgent action. Finally, because most of the activities to conserve biodiversity must be taken at the national level, the fifth catalytic action is to integrate biodiversity conservation into national planning processes.

Actions to conserve biodiversity in agricultural systems

The global strategy has outlined the thirteen guidelines for action to save, study, and use the earth's biotic wealth sustainably and equitably (World Resources Institute 1992a).

1. Eliminate agricultural policies that promote excess uniformity of crops and crop varieties that encourage the overuse of chemical fertilizer. Examples are input and

food price subsidies, overvalued exchange rates, research biased toward high-input agriculture, and credit policies discriminatory to minor crops.

2. Assert national sovereignty over genetic resources and regulate their collection. In doing so, ownership rights of physical and intellectual property should be recognized, and medicinal healers, small farmers, and other informal innovators should be included.

3. Strictly regulate the transfer of species and genetic resources and their release into the wild. Potential social impacts of transfers or releases should be assessed.

4. Establish incentives for effective and equitable private-sector plant breeding and research. To do so, intellectual property rights, plant breeders' rights, and farmers' rights should be recognized; public-sector institutions should be strengthened and private-sector incentives provided, which will help training provide a regulation structure; and national review boards for intellectual property rights should be set up.

5. Facilitate the exchange and development of technologies for conservation and sustainable use of biodiversity. To accomplish this, developing countries will require assistance from the international community, and information systems must be established or strengthened in developing countries.

6. Ensure that countries are free to decide whether to adopt intellectual property rights protection for genetic resources and how strong that protection should be. Mechanisms for resolving disputes related to intellectual property rights will profoundly affect the development, use, and conservation of biodiversity.

7. Reduce pressure on fragile ecosystems and wildlands by using land more efficiently and equitably. Skewed distribution of land ownership intensifies degradation of natural ecosystems.

8. Strengthen local capacity for maintaining and benefiting from crop and varietal diversity.

9. Promote recognition of the value of local knowledge and genetic resources, and affirm local peoples' rights.

10. Negotiate the collection of genetic resources on contractual or other agreements ensuring equitable returns, providing incentives to local participants through involvement and financial rewards.

11. Promote agricultural practices that conserve biodiversity. Promotion of crop diversification is an example.

12. Strengthen crop and livestock genetic resource conservation, and implement the proposed Global Initiative for the Security and Sustainable Use of Plant Genetic Resources.

13. Fill major gaps in the protection of plant genetic resources. Fruits, nuts, vegetables, and minor crops are examples.

The strategy emphasizes that new international funding, estimated at a minimum of $1 billion a year, is required to initiate the program outlined in the strategy. The continuing dialogue initiated in Rio between the developed and developing countries is considered important, and commitment at the national level will be critical. Within the context of national programs, the identification of program priorities and protected areas is assigned high priority, and the strategy urges prompt action and calls for steps to strengthen genetic resources conservation capacity and to build on the recent Keystone International Dialogue (Keystone Center 1991). Over the next 50 years or so when the globe's population will double, the most forbidding challenge in the implementation of the strategy will be to increase productivity of food, feed, and fiber on the same area of land and at the same time halt the decline of biodiversity.

Literature Cited

Arroyo, M. T. K., P. H. Raven, and J. Sarukhan. 1992. Biodiversity. In *Agenda of science for environment and development into the 21st century*, ed. J. C. I. Dooge, Gordon Goodman, J. W. M. La Riviere, Julia Marton-Lefevre, and Timothy O'Riordan, 205-19. New York: Cambridge University Press.

Barton, J. H. 1992. Biodiversity at Rio. *BioScience* 42(10): 773-76.

BOSTID. 1992. *Conserving biodiversity: A research agenda for development agencies*. Washington D.C.: National Academy Press.

Brady, N. C. 1988. International development and the protection of biological diversity. In *Biodiversity*, ed. E. O. Wilson, 409-20. Washington D.C.: National Academy Press.

Bruntland, G. H. 1992. Message from Gro Harlem Bruntland, prime minister of Norway. In *Agenda of science for environment and development into the 21st century*, ed. J. C. I. Dooge, Gordon Goodman, J. W. M. La Riviere, Julia Marton-Lefevre, and Timothy O'Riordan, 15-16. New York: Cambridge University Press.

CIMMYT. 1993. *CIMMYT in 1993: Poverty, the environment, and population growth*. Mexico City.

Dooge, J. C. I., Gordon Goodman, J. W. M. La Riviere, Julia Marton-Lefevre, and Timothy O'Riordan, eds. 1992. *Agenda of science for environment and development into the 21st century*. New York: Cambridge University.

Ehrlich, P. R., and E. O. Wilson. 1991. Biodiversity studies: Science and policy. *Science* 253:758-62.

Expert panels on biodiversity convention convened by UNEP. 1992. *Diversity* 8(4): 6.

Farnsworth, N. R. 1988. Screening plants for new medicines. In *Biodiversity*, ed. E. O. Wilson, 83-97. Washington D.C.: National Academy Press.

International Monetary Fund. 1993. *World economic outlook*. Washington D.C.

IUCN (International Union for the Conservation of Nature). 1980. *The world conservation strategy*. Gland, Switzerland.

IUCN. 1989. Draft articles prepared by IUCN for inclusion in a proposed convention on the conservation of biological diversity and for the establishment of a fund for that purpose. Geneva, Switzerland. Photocopy.

IUCN. 1991. *Caring for the Earth: A strategy for sustainable living*. Gland, Switzerland.

James, C. 1991. The transfer of proprietary agricultural biotechnology applications from the industrial countries to the developing ones: The international biotechnology collaboration program. *Rivista di Agricoltura Subtropicale e Tropicale* 1:5-24.

Kaul, A. K. 1989. Ethics of sustaining agricultural development in less endowed regions. Paper presented at the International Conference on Ethics and Environment at the Royal Swedish Academy of Science, Stockholm.

Kaul, A. K. 1990. Biological aspects of the sustainability of smallholder agricultural production systems in Indonesia and Bangladesh. *Proceedings of the International Symposium on Sustainable Agriculture*. Lexington: University of Kentucky.

OTA (U.S. Congress, Office of Technology Assessment). 1987. *Technologies to maintain biological diversity*. Washington D.C.: U.S. Government Printing Office.

Keystone Center. 1991. *Final consensus report: Global initiatives for the security and sustainable use of plant genetic resources*. Keystone, Colorado.

McNeely, J. A. 1988. *Economics and biological diversity: Executive summary and guidelines for using incentives*. Gland, Switzerland: International Union for the Conservation of Nature.

Miller, J. S., and S. J. Brewer. 1991. The discovery of medicines and forest conservation. Missouri Botanic Gardens. Typescript.

Miller, K. P., W. V. Reid, and C. V. Barber. 1992. Global biodiversity strategy and its significance for sustainable agriculture. In *Biodiversity: Implications for global food security*, ed. M. S. Swaminathan and S. Jana, 163-81. Madras, India: Macmillan India.

Raven, P. H. 1987. Biological resources and global stability. In *Evolution and coadaptation in biotic communities*, ed. S. Kawano, J. H. Connell, and T. Hideaka, 3-27. Tokyo: University of Tokyo Press.

Rodda, Annabel. 1991. *Women and the environment*. London: Zed Books.

Swaminathan Research Foundation. 1993. *Policy actions for biological diversity*. Proceedings no. 5. Madras, India.

U.S. Agency for International Development, Bureau for Science and Technology. 1986. Draft action plan on conserving biological diversity in developing countries. Washington D.C. Photocopy.

Wilkes, G. 1992. *Strategies for sustaining germplasm preservation, enhancement, and use*. Issues in Agriculture, no. 5. Washington D.C.: World Bank, CGIAR Secretariat.

Wilson, E. O. 1988. The current state of biological diversity. In *Biodiversity*, ed. E. O. Wilson. Washington D.C.: National Academy Press.

Wilson, E. O. 1989. Threats to biodiversity. *Scientific American* 261(3): 108.

World Commission on Environment and Development. 1987. *Our common future*. Oxford: Oxford University Press.

World Resources Institute. 1990. *World resources 1990-91*. Oxford: Oxford University Press.

World Resources Institute. 1992a. *Global biodiversity strategy*. Washington, D.C.

World Resources Institute. 1992b. *Global biodiversity strategy: Policy-maker's guide*. Washington, D.C.

Panel

JONATHAN TAYLOR
Booker Companies

I had to ring Norman Borlaug earlier this week about something else and I said I was going to be on a panel on biodiversity and could he give me any help. And he said, "Well, Jonathan, it has gotten to be quite complicated." Then there was a pause, "and it has gotten to be quite controversial." That's what I learned from Norman Borlaug. Perhaps one ought to stop there.

Of course, we all believe in biodiversity, except in my case for chickens—a subject I know a little about. Last night I heard someone saying that they also believed in biodiversity except for cows, and they certainly knew a great deal about cows—indeed more about cows than I knew about chickens. So, I think it's something we look at in the abstract and say, "Good and it must get on."

I think it is a bit more complicated than that. Perhaps, I can express it in more general terms. Imagine two boxes, both full of good things. The first box is about the need to feed the population—which is going to double over the next 40 or 50 years—and the need for higher animal and crop yields, the need to improve nutritional standards and standards of living, the need to deal with the discrepancy of population and wealth, the need to improve distribution of food, and so on; in other words, the need to make this burgeoning population better fed, richer, and happier. Then the other box is full of other good things, like the improving the environment, the amelioration of global warming, the preservation of tropical rainforests, the preservation of various ecosystems and of traditional cultures, and, last but not least, the maintenance of biodiversity.

I do see some conflict between these two boxes, and I think the imperative to increase food supplies may quite often conflict with the need to preserve plant and animal diversity. Certainly we do see heterogeneous, local crops being replaced by high yield, high-input, uniform varieties. We see peasant villages moving out of subsistence and into cash cropping and into agricultural labor. We see the domination of the global food system, whether we like it or not, by large companies and by governments. We do see a reduction in crop diversity, but often, on the other side of the scale, there is an increase in food productivity, in yield, and in absolute food production.

There may be answers to these two, sometimes conflicting, boxes. I believe that biotechnology may be a part of the answer. I think improvements in agricultural trade, the GATT system, the reduction in agricultural protection in

Europe, the United States, and, indeed, in Japan, can all be helpful steps. Somewhere population control has to come into it.

But there is a conflict. I think you will see it most clearly, probably, in China where food production will increase, standards of living will increase, but there will be more agricultural and industrial pollution. There may be a loss of biodiversity.

I think there is a meaningful role for an institution such as Winrock, grappling between what I see as a paradox of the conflicting priorities of these two boxes.

I will move from the sublime to the ridiculous, that is to say, from these large issues, which we can talk about, to little things that I sometimes know more about.

Chickens haven't been mentioned yet. Booker is the largest chicken breeding company in the world. One in three of the chickens you eat, wherever you are, will on average be produced from our genetic material. It's not an area where we feel a great need for biodiversity. We keep it within our own organization. We have about 30 pure lines, we use about six of them. We actually generate biodiversity by in-breeding.

In 1983, one of our breeding hens would, in 1 year, produce 130 chicks, which in turn would result in 176 kilos of broiler meat. Today, the progeny of one breeder will produce 267 kilos of meat. That's a 51 percent improvement in 10 years. Through hybridization and population genetics, productivity continues to improve all-around by several percentage points per annum.

We have to adjust to consumer tastes. In this country, the great need is for white meat because, for reasons that escape me, the American population has decided that only white meat is healthy. This is until we discover that dark meat is a cure for cancer of some sort, in which case it will reverse. Conversely, throughout almost the whole of Asia, there's a premium on dark meat. So, we have a certain amount to do to satisfy the consumer.

We would love to see a great gene bank for chickens. I don't think we would see the need to pay for it. It would be welcome if it was there. But we're fairly happy with what we have, and we do have a lot of diversity in-house.

Let me say a word about salmon farming. When we started out salmon farming, there was a high-minded debate about the effects of salmon farming on pollution of the seabed from the various detritus and feces of salmon. There was a lot about visual pollution, mainly from people who had holiday homes overlooking lochs, who felt these cages wouldn't improve the view. In the United States, you have the same problem in Puget Sound. Then there were more theoretical questions about what farmed salmon would do to the wild populations. What would happen if domesticated escapees crossbred with wild fish? Maybe the salmon will lose its navigational facilities, and all sorts of terrible things would happen. Maybe farming would introduce diseases and interrupt wild runs. In any event, the diseases come from the wild fish. There hasn't, I don't think, been any problem with the escapees and the genetic process. And the last person to pollute the seabed would be a farmer because his livelihood depends on the sustainability of his farming activity.

I could have seen a meeting like this discussing these issues in a serious way and perhaps coming to the conclusion that we shouldn't tamper with salmon. The reality is that the price of salmon has halved in real terms over the last 10 years. Ninety percent of all salmon consumed is farmed. Consumption is rising at 15 to 20 percent a year. That, I think, is the real world, and we must be careful that we do not inhibit progress.

Finally, a word about environment as seen from a corporate perspective. We are a company that devotes considerable attention to the environment, but it's a happy environment because in our terms everything we do about the environment—from making everyone drive diesel cars to new forms of packaging, recycled packaging, and reduced energy use—all save money. In our terms, there is no conflict between good business and the environment.

I am sort of a Darwinian. I think we're going to survive and we're going to make progress. If there is a choice, a purely theoretical one, between preserving the dinosaur in its natural habitat and providing the Turkana woman with a continuing livelihood and a Sony wireless set, I know where my priorities lie.

W. RONNIE COFFMAN
Cornell University

I want to start with a statement that I heard this morning: "The reason we don't have biodiversity in agriculture is that we have already done away with it." In a way, what Jonathan said about chickens illustrates that the audience expected, I think, somebody to take issue with it. But that's essentially correct. At least it's correct in rice, which I know something about.

Norman Borlaug was mentioned earlier. If you stand around in a field with him for a while, he'll tell you about the origin of agriculture and how it was started by some women who had lousy hunters for mates. Of course, back then you had the maximum in biodiversity, but those women soon found out that they needed to focus. They needed to increase their productivity, and we've had that going on for 10,000 years. We're down to just a few crops as we all know. Some people say about 20. But if you think about the ones that really do the job, it's just a few cereals that account for 70 or 80 percent of the food. Rice, the most important one, is the primary staple for half the world's population.

I'll relate one instance to illustrate where we are—the situation with cytoplasm in rice. As you know, all organisms are composed of cells. You have the nucleus, which contains most of the genetic material, and you have the cytoplasm, which is the rest of the material in the cell. So, when you make a cross, half of the nuclear material comes from the plant that contributes the pollen, the male, and the other half comes from the plant that receives the pollen, the female. But all the cytoplasm comes from the female, always. IR8 was the first improved variety from IRRI and it came from a cross between Peta and Dee-geo-woo-gen. Peta was the female so it contributed the cytoplasm. Peta came from a cross between the varieties Tjina and Latisail. That's as far back as we know. Tjina was the female, so the Tjina cytoplasm contributed the cytoplasm of IR8, and then IR8 was used in the parentage of nearly all other improved varieties that are deployed all over the rice-growing world.

You may say, well, that's all right, there must have been other parents that were used as females. But actually what happened was, plant breeders preferred to sit down when they were making crosses, as opposed to standing up, so they always used the shorter one as the female. So, essentially all the deployed rice varieties all over the world, except for a few that have been produced lately, have Tjina cytoplasm.

So, there are potential problems. It's the worst possible situation until you consider the alternative and the alternative wouldn't be pretty. We know that the real price of rice today is half of what it was when IRRI was established. If the price went back up to the original level, we can only imagine what kinds of problems we would have.

ROBERT O. BLAKE
Committee on Agricultural Sustainability for Developing Countries

I would like to respond to a question I have been asked about the World Resources Institute's work on a biodiversity strategy, put together in anticipation of the Rio conference. People in almost every country were involved, and not just people that you would normally think should be involved. There were government officials, there were people from the business community, there were people from every walk of life. In the end, even more important than the document was the process of bringing people together to think about the problems of preserving biodiversity. An attempt was even made to get ordinary farmers and ordinary producers involved. That wasn't as easy. In the Committee on Agricultural Sustainability for Developing Countries we, too, make a special attempt to get a wide range of interests involved. For example, we have recruited women's groups and religious groups as members in addition to environmental groups and NGOs. The times and the complexity of solving agricultural problems demand it. I think that this is reflected in the group that you've assembled here. You probably wouldn't have brought together the same kind of a group 10 years ago. I urge Winrock to continue to reach out to new groups and to make it one of its most important goals.

Just a couple of comments about ex-situ and in-situ conservation discussed earlier: One interesting development on in-situ conservation in the international agricultural research centers is the concept of heritage sites—finding and then protecting sites where the wild relatives of major crops are located so that the evolution of the genetic base of these crops can continue.

The question of access to genetic resources is coming into sharper focus. Some interesting work is being done in Costa Rica. Costa Ricans are making available their genetic resources to companies in return for financial help in identification of genetic resources. Costa Rica is not relying just on the protected areas of the country (essentially the national parks), they're also setting up buffer zones around protected areas and giving farmers, hunters, and people who live nearby—and who too often abuse those areas—a stake in protecting them.

In the United States, some fascinating work is going on in planning how to go from protecting biodiversity on a species-by-species basis to protecting whole ecological areas. Secretary of Interior Bruce Babbitt feels that the only way to protect endangered species is to protect larger areas of habitat for plants and animals. This has to be done, again, by giving people around those areas an interest in having this preservation.

The ex-situ conservation work of the CGIAR is important. In my view, the chances of developing nations doing much about this in the short run are small. They simply don't have the resources. There are many very good people, struggling hard to get those resources, but they're probably going to have them in the near future. For as far ahead as you can see, the CGIAR system work on this is absolutely crucial.

I've been surprised to count how many people in this group here have or have had connections with one or another of the CGIAR centers. You're the moral majority and remember that, the moral part, too, because the CGIAR system is in danger. We have to face up to the fact that every day the system is being asked to do more and more, as our horizons expand and as the urgent need for research results becomes clear. At the same time, the money is disappearing.

This year USAID had to cut its funding of the centers. Next year, it will cut more. Other donors will follow along. We've been talking in the Committee on Agricultural Sustainability for Developing Countries about how we can reverse that, how we can convince the World Bank, perhaps, to put more of its earnings into the CGIAR. But if we're not careful, we will find that the research we all care about will be chiseled away at the edges in a very dangerous way.

Now I'll give you the bad news: The system itself is not generating the demand that this be changed. The system itself is being passive in the face of this. Groups like our own and others will, from the outside, help. But from the inside, we need to hear a demand that says: "You're no longer cutting fat, you're cutting into the bone and into the meat. We have the capacity to do more if we can get the additional funds, and if we don't get funding, this is exactly what's going to happen."

So I ask every one of you not only to recognize this, but to talk this issue up with your groups, and begin to generate the kind of action that's needed. A relatively small amount of money produces high-quality results; this has to be a priority. We have 15 years between when some of these research efforts are started and when the results are used by a substantial number of farmers. We have to take action right now.

Discussion

Kaul: When I visited ICRISAT recently, I heard several examples illustrating the need for ex-situ collections. They told me that Somalia, during the civil war, lost most of the materials it had in the germplasm bank. Farmers ate all their seed when the country was facing a crisis. Luckily, germplasm of 300 varieties of Somali sorghum had been collected earlier and carried out of the country, so it could be repatriated. Comparable cases were cited for Ethiopia. There was similar concern expressed about Afghanistan where a tremendous repatriation of germplasm has to occur after things settle down there. They also mentioned Central Asia where it was a different story. There monocrops of cotton were grown on an area the size of Texas and Arkansas put together. So they lost the germplasm base of most other crops, which farmers now urgently need. And not only did they lose the material, they also lost the techniques of production. They also lost the biota, along with it and a whole system of production.

So, I couldn't agree more that it's an urgent issue. Something has to be done because on the very day I visited ICRISAT they were told that their budget had been slashed by a couple of million dollars. I wonder if the community like ours should not be seriously doing something about it.

Harwood: A third of the world's poultry germplasm is apparently coming from a few genetic lines. The resultant high-intensity of production has led not only to high productivity, but to high efficiency of input use. The response of these improved breeds to good health care and nutrition has been outstanding. As a result, however, chickens have disappeared from villages because small farmers could not compete with vertical integration and large-scale, capital-intensive methods.

You could argue that the cheap poultry meat has been a tremendous social good. But I would like to ask Jonathan Taylor, if he doesn't worry about the instability of being dependent on quarantine, exclusion of disease, and perhaps prophylactic use of antibiotics to keep those animals healthy, in a narrow genetic base? It has been a tremendous success, but the downside of that, were it to crash, is disastrous.

Taylor: I agree with almost all you say. But I don't think the genetic base is quite that narrow. We're using six lines actively in current programs, and we occasionally go back and draw from our gene pool for various characteristics—feather sexing or whatever.

The disease question is quite interesting in poultry. We devoted an enormous effort in the 1970s to breed resistance to chicken cancers. And then just as we got there, an effective vaccine was found, and we were set back for about 6 years as a result because we had been sacrificing productivity characteristics to resistance.

I think there is enough diversity. Particularly in Southeast Asia, chicken is the cheapest animal protein. It's the easiest for a small producer to handle and the productivity characteristics the breeder tries to bring out are relevant whether it's a backyard operation or an industrial operation.

I think the progress of refrigeration goes hand in glove with the progress of the chicken. But it is the first step in a varied diet that introduces animal proteins on top of traditional carbohydrates. It has a lovely developmental impact.

Your horror scenario of something going fundamentally wrong, a disease maybe, or a breakdown of some sort in the genetic cocktail, I suppose is always there, but I think we do have a lot of diversity in the gene pools we're keeping.

Paarlberg: I would like to ask if it's fair to distinguish between different degrees of severity of species loss or biodiversity loss. The most severe might be outright extinction of species. A slightly less severe condition would be loss of diversity within the species, not yet extinction but loss of some of the genetic base—getting down to only two California condors, perhaps. That's a loss of biodiversity, but the species is still alive. A third condition would be diversity loss in the farmer's field, where monocultures are being produced even though there may be a large array of genetic materials still available in heritage areas or in gene banks somewhere. Using such categories, are we in "condition 1" in agriculture or are we closer to "condition 2" or maybe "condition 3"? And if we're in condition 3, how much do we have to worry?

James: The problem is that we know so little about biodiversity. Wilson from Harvard suggests that we may be out by an order of magnitude when we say we know 10 percent of the species. So what we need is information on what the species are and on the importance of different species within that ecosystem. We don't know the function of different species, and almost certainly some are more important than others. Until you get that information, you can't make a good judgment.

Coffman: We know in general with a crop like rice that the germplasm collections are relatively good. Most of the Chinese germplasm was lost during the cultural revolution, but it was fairly well collected in other parts of the rice-growing world. The thing is that the deployed germplasm—and it's difficult to get hard data—is relatively uniform. It all stems from the work done at IRRI in the early 1960s and 1970s. So we're not in any of those conditions that Rob Paarlberg describes, unless something happens to the rice crop. There's no reason to think something will happen, but if it does, it will be the mother of all type one errors. We don't have any real reason to believe that there is a problem except the fact that the deployed varieties are fairly uniform, and Murphy's law is always there.

Kaul: All biotechnology still is dependent on the existing variability, which has to be shuffled, reshuffled, retailored, and other things. The question is not how much we have lost in what category; the question is are we willing to lose any variability at all?

I have worked with the legume *Lathyrus sativus*, which is one of the very few truly drought-resistant legumes. It can be grown under very arid conditions. Its only problem is a freak amino acid that causes paralysis if you eat it. Through incremental breeding procedures, we have almost eliminated that factor from this crop. Can you imagine, when it is absolutely toxin free, what a revolution that can bring in areas that don't have water? So, I don't think we can put a price on any single gene, much less a species.

I don't think we should label variability for the sake of how we can perceive its use today, or 10 days from now, or 10 years from now. Who knows when we would need what particular organism for what purpose?

Falcon: It seems to me if we have to wait to do anything in biodiversity until we have the ecology of all the species mapped, we're in bad trouble because the world may be gone by then.

My question has to do with patenting. It seems to me that one can distinguish three positions. One is that the patenting process is absolutely critical to generate investment. Another position you might find in India, where the idea of patenting a life process is a moral or philosophical anathema. Other Third World countries and some of the international centers hold a third position: that because much of the genetic material comes from the Third World, how can we assure that the patenting process doesn't deprive Third World countries of the results? I believe that's a serious problem and we're stuck dead center on it. Does the panel believe that, too? Or are we making progress?

Horsch: Patenting is not essential for investment, it's only essential for private investment. It's perfectly legitimate to go for public investment and to exclude or to discourage private investment. That's a societal choice. My recommendation would be that society is better off, by a long margin, to stimulate private investment. As for access, a patent has a life span of 17 years, most of which is eaten up before the products even reach the market. For most products you're looking at, if you're lucky, that is 5 or 6 years of protection.

The information that went into the discovery and the ability to replicate the discovery is available publicly at the time of issuance of the patent.

James: In ISAAA, we get a donation, which means waiving of the patent. I can see the evolution of this process where you may be able to get access for a developing country through a subsidy. There are many joint venture possibilities available at that stage. If, in fact, you have the primary germplasm coming from that country, then there may be some recognition for the value of that product. And that's the way that I can see it evolving with patents.

Taylor: It is a difficult question, but it doesn't apply to all parts of genetics. If you're dealing with hybrids, the question of patents is not too material. I think a breeder who is achieving year-on-year incremental performance improvements is much more interested in holding that competitive edge than in the protection of plant breeders' rights, or whatever, because his 1993 model will always be that much better than his 1992 model. That's in the end what will sell his product. But for people dealing in pure lines, and certainly for biotechnology, proprietary rights are extremely important.

Blake: The people in the U.S. Trade Representative's office say there's a good chance that this round of the GATT will produce a lot more in the way of patent protection. They are less worried about the protection of agricultural patents than they are about the pirating of rock music and things like that. It's not as easy to pirate these other things.

Kaul: The conflict over patents is not going to be easily solved at GATT because suddenly people have become aware that they own something that they previously didn't realize was valuable. It was robbed, they say, or it was taken away. Now they want to guard it and ask for compensation.

Peterson: It seems that there's an idea that biotechnology is going to be the next revolution, and through it we will be able to feed the doubling of population over the next 30 or 40 years. If biotechnology is the wave of the future, will it change the way we do research? And, if so, does that mean I have to change the way I invest in research capacity, i.e., is this going to cost me more money, the same amount of money, or less money? Or is it going to make me invest in laboratory test tubes rather than tractors?

Horsch: I agree that this will be an evolution rather than a revolution. If you look at what's in field tests today—what was started in 1987 that now looks promising in 1993, there's nothing that will save the world by 2000 or 2010. We can hope that we will make discoveries that aren't in field tests or greenhouses today, but that's just a hope. So, I think it would be irresponsible to say technology is going to solve the future problems, and we can lie back and not worry about it. You're still going to need irrigation. You're still going to need mechanization. You're still going to need breeding.

I think we are facing diminishing returns. Research will get more expensive. Biotechnology research is certainly more expensive. So, I would anticipate larger investments being needed. In the past, we have gotten a huge bargain from plant science research. We might have to pay a cost more in line with the true value in the future.

James: I don't see any quantum increases in production resulting from the technology that's available in the near term. On the other hand, there are technologies around that could be used in the developing countries within the next 5 years that can make a substantial contribution to the environment. The lead products that are likely to be adopted in the developing countries within the next 5 years are in the area of plant protection, which provides a much more environmentally sound way to control diseases. Two-thirds of cotton plantings are in the developing world, and there is a very large amount of investment in insecticides, along with environmental pollution.

Coffman: It's important to look at where the investments are being made. It's fair to say that most of them are being made in the private sector and not to meet the food needs of people. Most of the investment is being made in the private sector with the hope of some return. There's nothing wrong with that, except you have to imagine what the impact may be on important export crops for tropical countries. We know that hunger is caused by poverty, not by lack of food being produced. So you can see the real possibility that major export commodities from the tropical countries are going to be lost to new technologies that allow them to be produced in the developed world. Investments are being made in that area.

Paarlberg: I'm worried that if biotechnology investments remain a private-sector activity, we won't level the playing field between temperate-zone and tropical-zone farmers. It will tilt it further if the private sector invests first in temperate-zone agriculture, second in high-potential farming in the tropics, and then, third, or not at all, in low-potential farming in the tropics. If you don't get public-sector involvement here, you will just see more spread in the performance potential of these three already very different groups of people.

Horsch: I interpret that as a call to action like the one Bob Blake left us with.

8
Round Table Discussion: Implications for Development Assistance

Introduction

ROBERT D. HAVENER
Winrock International

This is the wrap-up session of what, in my view, has been an extraordinarily thoughtful, interesting, and provocative set of discussions around a set of extremely important issues: the extent to which our soil and water resource bases are being severely degraded at this point and whether they will have the capacity to provide the food and fiber that we will need into the first quarter of the 21st century and beyond. Many of us feel that the next 30 years is an extremely critical period. The 3 billion people who will join us on planet earth between now and 2025 must, by and large, be accommodated in densely populated and poor parts of the world. The extent to which they can contribute to improved human welfare and the advancement of mankind versus the extent to which they will be irreparably underserved by the legacy that we leave them have been the big questions that we have been addressing.

We've looked at soil and water resources; we've looked at poor people, human resources, and the environment and how, and to what degree, they interact. We've looked at various production systems—systems of livestock production, crop production, forestry production—and how these impact on the environment and on the people that occupy the environments under threat. We've looked at genetic conservation as a duty—preserving the genetic treasure trove that we have and the biodiversity that is at our disposal for the use of future generations. All these extremely important questions have been addressed by knowledgeable observers. The discussion has been first class.

We've asked Vernon Ruttan to pull together the big ideas that have come from that process, to address the major issues in development assistance, and to share with us how we might proceed in directing our attention toward some of them.

Major Issues for Discussion

VERNON W. RUTTAN
University of Minnesota

Five visions

In considering the issue of development assistance, let me remind you that in the period that we've been involved in this business we have been guided by a series of five visions. The first vision emerged during World War II—a vision of a liberal political and economic order after the war. It was this vision that gave rise to organizations like the International Monetary Fund, the World Bank, the United Nations, the trade organization GATT, and others. That vision has informed U.S. policy throughout most of the period since World War II. The second vision was the vision of technical assistance, articulated by President Truman in his Point 4 address—his inaugural address. Truman proposed bringing the benefits of U.S. science and technology to the developing world. Third was the vision that President Kennedy articulated during his campaign and during his short period in office that a more prosperous world would be a safer world. He turned around the security concerns that had dominated the middle of the previous administration and focused it on development. Fourth, there was the vision of the members of the House Foreign Affairs Committee in the early 1970s—the new directions vision that focused on meeting basic human needs and on human rights. Fifth, there was the linkage between economic and security assistance in the first Reagan administration, the landmarks of which are the Carlucci Commission Report and several speeches by Alexander Hague and Jeanne Kirkpatrick. I'll come back to the issue of vision, but it's sobering that, although each of these visions has held some currency, the half life of a wave of the future has not been very long.

Three concerns

Let me now outline three concerns about the sustainability issues that we have been talking about. The first is, what do we do as an encore to the Green Revolution? Let me remind you that if you walked around the world in 1900, you would find grain yields, in the better areas, of about 1,000 kg/ha, plus or minus a couple of hundred kilos. If you went around the world today and looked at grain yields under the best conditions, you would find a rice, wheat, and maize yields of around 8,000 kg/ha, plus or minus a couple of thousand kilos.

But we are now approaching a new ceiling, somewhat like that 1,000-kilogram ceiling that existed before we began to apply Mendelian genetics to plant breeding. In the United States, maize yields have risen 130 kg/ha annually since I left the farm in 1940 when maize yields were 2,000 kg/ha on the average. They are now 8,500 kg/ha. Yields are still going up 130 kg/ha per year, but that is a much smaller percentage increase than when yields were 2,000 kg/ha. If you plot it on a log chart, which shows percentage change rather than an arithmetic progression, it looks like an S-shaped curve. We have already heard about the difficulty of raising rice yields in maximum yield trials at IRRI.

There has been a certain amount of discussion about cheap food policy, and I would point out that there are two kinds of cheap food policies. One is the type of cheap food policy that Thailand followed when it put a 30 or 40 percent export tax on its rice. That's a cheap food policy that both depresses farmers' income and dampens production. But there's also the cheap food policy based on technical change that is a win-win situation. It permits increases in farmers' income and reductions in consumer prices.

Since 1850 wheat prices have declined continuously to $150 per metric ton. In today's terms, wheat prices would be above $500 per metric ton if they were at 1850 prices. Rice prices have declined only since mid-century. I would argue that that decline in wheat prices gave a tremendous advantage to those parts of the world in which wheat was the basic subsistence commodity as compared with the parts of the world in which rice was the basic subsistence commodity. I don't think it's entirely an accident that industrial progress and income growth were higher between 1850 and 1950—until rice prices began to decline—in the parts of the world that grew wheat and consumed wheat as the basic subsistence good as compared with those areas that consumed rice.

I will view it as a failure if we do not continue to lower the real prices of basic foods. When somebody tells me that their technology needs to have higher prices in order to be adopted, I take that as a failure also.

The second concern is with the spill-over effects of agricultural and industrial intensification. We have been talking a great deal about the resource constraints on growth and the role of technology in overcoming those resource constraints. I think we know how to remove the technological constraints. I'm not as sure that we will make the investments to do it. But it seems to me that the most serious constraints today are emerging as the by-products of agricultural and industrial intensification. From the agricultural side, it is groundwater pollution and soil erosion. From the industrial side, it is acid rain, global climate change, heavy metals in the soil, etc. We will have to attack that problem with at least with the vigor that we have attacked the problem of applying technical change directly to agriculture production.

It may seem a odd to mention health constraints as my third concern. But in most poor countries, the little data we have indicates that farm people lose 10 to 20 days a year from illness. That doesn't mean that they're not ill more than that. It simply means that there are that many days when they aren't able to get up and go to work. When you begin to consider, in addition, the potential implication of malaria resurgence, of tuberculosis resurgence, of the failure to make progress on parasitic diseases (with a couple of exceptions), of the high cost of dealing with infectious disease, of the health effects of environmental change, and of AIDS, it's not too difficult to envision the emergence of a global health crisis in the early decades of the next century.

At a recent conference, Don Henderson, formerly dean of public health at Johns Hopkins, who led the global smallpox eradication campaign and who is now an assistant secretary at the Department of Health and Human Services, drew a comparison between agriculture and health. He said, "In agriculture you have an institutionalized extension service that helps farmers understand their production functions. In the health area we have not yet institutionalized a system that enables families to understand how to live more healthy lives. The developing countries will not be able to afford the hospital-based systems that we have, and we can't afford them any more either."

Research implications

I would like to talk a bit about the research implications of these concerns. As I look at the world today, I see a series of island empires. (I draw on Jean Mayer for that analogy.) We have an island empire of crop agriculture. We have an island empire of livestock agriculture and veterinary medicine. We have an island empire of environmental research. And we have the island empires in the health area. There are few bridges between these islands. We're going to have to face, as we build irrigation systems, as we intensify agriculture, building more bridges at least at the operational level.

I would like to say a little about the vision that has emerged or is emerging in each of these areas. In the agricultural area, we have a fairly clear vision of what a global agricultural research system should look like that if it is to be fully articulated with the producer. When we began this development business in the early 1950s, we thought of the extension worker as the client for agricultural research. We thought of the peasant as being the major obstacle to agricultural development. I remember that when Ben Higgins, one of the early development economists working in Indonesia was asked, "What is the major problem with Indonesia?" His response was, "Too many peasants." Ted Schultz turned us around on this when he insisted that we regard the peasant as poor but efficient, operating in a society that had not made available the tools needed to achieve productivity. We have a vision of an international agricultural research system that is funded by the CGIAR, which has a Technical Advisory Committee (TAC) that looks at the strategic priorities, that focuses on how to allocate and reallocate resources—not that they have enough resources. And we have a vision of national agricultural research systems that are articulated with the international system. This international system will never fill the gaps until we get strong national agricultural systems. But if we had a world of strong national agricultural systems, we would still have to invent the international system to facilitate the communication and the priority setting among them.

We also have a vision of how to achieve articulation between public and private research systems and the changing role between the public and private sector. And we understand how to articulate both of those systems with producers.

We have the vision. I'm not saying that the system is fully in place. The formerly centrally planned economies are not part of that system. They should be. It should be a global system and not simply focused on the developing countries.

In the health area, we're just beginning to have such a vision. We only have two international institutes. We have the Diarrheal Center in Bangladesh, which has been very productive. It has developed an oral rehydration procedure that has saved millions of children's lives. We have an International Center on Insect Physiology and Ecology that does some work on the insects that are relevant to human health but mostly on agriculture. Probably we will not invent a series of international institutes in the health area. We will probably depend upon networks because, at least at present, the international environment for raising funds is too constrained, but I believe we need to have in health what TAC gives agriculture, that is, an international structure that can think about how to approach the global health issues.

We do have, in a few countries, the beginnings of systems that focus on the family as the primary customer for knowledge and materials about health. In western Kenya, for example, there is a system in which mothers are viewed as the primary deliverers of health services and in which an effort is made to make available knowledge, materials, and technical assistance to enable mothers to perform that role. The International Commission on Health Research for Development has begun to articulate such a vision.

In the environmental area, it seems to me, we have only begun to articulate the vision of a global research system. I was delighted to see a chart of the structure in the genetic resources area based on the World Resources Institute report on biodiversity. It went all the way from the farmer to the international system. We don't have that with respect to global climate change or the major global resource areas. My sense, when I go to the Board on Global Climate Change Research at the National Research Council, is that the research at the international and national levels is directed at the negotiators of international treaties. We haven't gotten far beyond that. We haven't thought of the linkages that will have to be built all the way down to the decision makers and of the essential national research that has to be done.

In the health area, when I first came back from IRRI, I started talking to the health research people about building national research systems. The response was that it can be done in the former colonial medical research institutes, which are in developed countries. But we know that it doesn't get done. Only 5 percent of health research is on the health issues that are of most significance to developing countries. The situation in the environmental area is comparably weak.

Future of development assistance

Let me return to the issue of vision with respect to development assistance. Nothing has yet replaced the Cold War as a device for mobilizing assistance to developing countries. Every president from Truman through Reagan, when it came down to the budget issue, could always argue, "If you don't come across with the USAID budget, the Russians will get you!" That period has ended. In Washington there has been an effort find something new to be scared of so that the resources it will take to sustain the USAID budget can be mobilized.

The USAID budgets have been declining in real terms since 1985. It would nice if sustainability could be the mobilizing force. But I don't expect that it will. The prospect is for a continued decline of the USAID budget in real terms and continuous efforts by the aid constituencies—agriculture, environment, population—to carve out a larger slice of a declining USAID budget.

That is not what I hope will happen. I hope that Ambassador Blake can help mobilize the constituencies needed to support agricultural research. I believe that we will not remobilize the constituencies until we are able ourselves to articulate the kind of a world we want to live in the 21st century. That was what came out of World War II. We were able to articulate, as we built those international institutions, the kind of a world we wanted to live in. By articulating it at the highest levels and incorporating that articulation into the political ethos, we could move ahead to advance the more specific agendas.

I can't articulate that vision. But I can suggest a few components. It is essential that Russia reemerge as a major economic and political power capable of cooperating with the United States in establishing a world order. If that doesn't happen, the other things that I would like to happen won't happen either. We need to see the emergence of a world without refugees. Every decade the number of refugees has doubled. We are confronted with the fact that nations and states are not the same, and their borders are not the same. We are confronted with a nation of Kurds living in at least three states. We are confronted with the Balkans. We have nations within states that do not want to be in the same state with their neighbors. At the same time, we know that they cannot be economically viable as states built around their own ethnic community. That means we have in this world a constitutional problem: how to build constitutions that facilitate sufficient autonomy of nations and sufficient openness of economic systems to be viable. We are now facing that in Minnesota. Are we going to live up to our treaty rights with the Chippewa and allow them to spear where I fish? That's a big issue in the state legislature today.

Perspective

It does seem that we have begun to confront the challenge of this century in agriculture. That challenge has been to make the transition in one century from a resource-based agriculture, in which raw resources accounted for most of the production growth—when almost the only way to get growth was to expand the area cultivated—to a science-based agriculture in which the resource base plays an important but diminishing role. We will have to make that transition feasible for every country in the world if farmers are going to live up to the expectations that their societies place on them.

I believe that most of the growth in the world's agricultural production, from now on, will have to come from the robust soil areas that are already providing most of the agricultural production. Filling in around the edges in some of the resource poor areas or the fragile resource areas is probably more important to the people who live in those areas than it is to the rest of us. We need to do it. But, in my judgment, that's not where we're going to get the food supplies of the future.

Round Table Discussion

ROBERT BERG
International Development Conference

As I see it, global environmental issues are on the agenda for this country. Global agricultural issues are not on the agenda. That came home to me when Tony Lake, President Clinton's national security advisor, gave his first speech to a group of mostly African ambassadors. He explained that there was a great concern about sustainable development. He talked about USAID's work in environment. And he said, of course, the underlying issue is population and AIDS. And I thought, for an Africa analysis not to mention agriculture or for a sustainable development analysis not to mention agriculture, that meant a little education is needed.

The basic issue in global agriculture in the next century is that, as we said when we were teenagers, "we're cruisin' for a bruisin'." Is the world up to this? The United States has four major ways that it can respond. It is giving blurred focus to one and little focus to three others.

The way that it is generally mobilizing support is through its bilateral assistance programs, which are focused on sustainable development, but not clearly. USAID's future mandate will include sustainable development, consolidation of democracies, and some conflict resolution. But there is a poor focus on these issues—a narrow concept of bilateral assistance in which dealing with declining resources is more a preoccupation than developing opportunities for leadership on global issues.

The human resource development underpinnings of our strategies will be maintained, I think, but they're not well-focused and there's a considerable danger that we will neglect our international education resources in the next few years. A major problem will be that, out of concern for operating expenses, the bilateral program now covering over a hundred countries may be downsized to as few as 50 country programs. The brunt of that will be in Africa where, we have all agreed, the greatest food and sustainable development crisis exists.

The second challenge to be faced, and one that is not on the agenda for the United States, is how to take its largest international assistance program, its Middle East program, and shift that to the real insecurities of the area, which I would submit are water security and finding a viable economic role for a future Palestine. Instead, as happens with all major donors, the larger the program, the less the focus. I think that we have to see that that's on the agenda, and Winrock, I think, could have some important things to say on water security there.

The third challenge is the U.S. role in the multilateral development banks. The World Bank's rhetoric is excellent with outstanding people like Ismail Serageldin, the new sustainable development vice-president, and his colleague, Mohamed el-Ashry. But the question is not rhetoric now, it is real performance and the question of whether we can sort out loyalties—whether one has to be loyal to the state or whether we can shift loyalties to people and their sustainable development.

And the fourth challenge is the United Nations system. Clearly, there has been a sorting out of boxes, but the content of those boxes hasn't been figured out yet. I think the content is up for grabs and the United States can have a major role. The bright spot is going to be UNDP, where, with Gus Speth in charge, we can see UNDP becoming what it was meant to be, which is the systems leader in the United Nations system. Sustainable development and the Global Environment Facility could have a real boost.

The dull spot clearly is FAO. I think it is important to politically orchestrate a post-Saouma regime, which can bring respectability and leadership potential to that institution. There are some good patterns that have been set. The International Fund for Agriculture Development faced its criticism and with the help of people like Elmer Staats put its reputation in order. Even Unesco, another weak sister in the organization, has shaped up, a much underappreciated fact in the United States.

International leadership is assured on a variety of issues. There's momentum in health. There's momentum in education—as seen at the Jomtiem meeting. There's momentum now in environment from the Earth Summit in Rio. The summit for children made that issue politically respectable. Political leadership is expected at the 1994 United Nations conference on population, at the 1995 United Nations conference on women and development, and at the very significant 1995 world summit on social development. But there is no international momentum now in food and hunger.

So a real challenge for the global community in the food and agriculture area, and I submit for Winrock, is how to make the food and hunger issue politically urgent and respectable for senior political attention. Where will the back room for that be? There has been a back room in all the great movements, e.g., UNICEF for children, the Highlander Center for civil rights. Where is the think tank going to be that advises key leaders on what the opportunities are, and how good sense and good politics can come together on food and hunger issues? And how will the Clinton administration understand that it alone, of all political centers, national or international, has the opportunity for leadership on this issue?

Here's an opportunity not only for rekindling the food and hunger issue, but fundamentally for moving away from the sterile structural adjustment strategy of the 1980s to something of enduring worth to human existence, building, if you will, on the 1980s but moving well beyond it. Certainly one would hope that the Clinton administration could see the opportunity of using the next World Bank/IMF Board for announcing something beyond structural adjustment.

As I think about these things, obviously this institution comes in mind.

ROBERT O. BLAKE
Committee on Agricultural Sustainability for Developing Countries

I want to say a word about what the Committee on Agricultural Sustainability for Developing Countries does. We work mainly on trying to influence the World Bank, USAID, and the Inter-American Development Bank. We have also worked a good deal with the CGIAR institutions, the international agriculture research centers, because nobody else is paying much attention to them. Frankly, it has been rather tough to get any real attention to agricultural development. With most people, agriculture draws a big yawn—for reasons that we don't need to go into today. But that is the fact. Then there's the deeper question that Vern Ruttan and Bob Berg brought up: Just where does foreign assistance fit in with our national priorities today?

It's clear to me after having a chance to talk with the people of the Clinton administration, who are beginning to articulate what they think, that they basically share our concerns about sustainability of agricultural development. I don't have any question about that. But they also believe that politically this is not the time to emphasize this, because they have higher priorities. They have said that, for the United States and for the world, getting the U.S. economy back into a position that it can make a contribution commensurate with what people expect of us is about as important as anything else we can do. I can't quarrel with that. I can tell you that it's almost frightening how much people depend on the United States: How much they depend on us for leadership in defense—nothing happens in Bosnia without us—how much they depend on us for scientific and technical leadership. Nobody else is providing this kind of leadership. We have to do it.

Another important U.S. priority is how do you make Russia part of the democratic world and a free economy? Those are the administration's highest priorities for the near future, and they are correct ones. But in themselves, they are not enough. It will be quite important to begin to get a broader and longer range articulation of our interests that says that anything that is so fundamental that it concerns 80 percent of the world's population has to concern the United States, anything that is so essential to the welfare of those people as feeding them, anything that is as essential as bringing about a 50 percent increase in the world's food supply within 20 years—unheard of in history, I believe—has to concern our government and has to concern all Americans.

How do you get the American people to accept this priority? Over time we've always gotten there, as Vern Ruttan said, by a combination of vision and good luck, and, I would add, very often by scaring the hell out of people. That was what the Cold War did. That's just the way we are as Americans. But we must also support a strong idealistic vision. It's what allows our humanitarian instincts to surface.

How do we position ourselves, particularly on sustainable agriculture, until the inevitable day that it comes to be broadly recognized as a high priority? First, we clearly articulate the problems we face. Say it again and again. Even if nobody listens, say it again and hear the echo coming back. Second, we have to make careful choices in agricultural research. What research will have the biggest payoff in productivity for the countries that are faced with the biggest problems? We must also develop a better capacity to influence positively the World Bank and UNDP. They are beginning to change. But some people call this the "dance of the elephants" because it happens in such a measured way. Yes, the culture of the World Bank is changing, and we can help.

I would like to end by saying a word about what I think Winrock can do. First, there's going to be a gap between needs and the funds to support them. You will be hard pressed to maintain the kind of research for which you previously received U.S. support. We'll work to help you get that support. But I think you have to depend a lot on the men and women here in Arkansas. Their arguments will count a lot with the Clinton administration. Second how can you better articulate the kind of vision that was talked about here? How do you use your incredible intellectual influence and show how agriculture, particularly agricultural research, fits in? On another subject, I would ask you to try to define, in very practical terms, what sustainability means in agriculture, what it means in growing trees, what it means with cattle, what it means in every aspect of agriculture at home and abroad. And last, and this may be the most important, get somebody as good as Bob Havener as your leader.

E. WALTER COWARD, JR
Ford Foundation

I take it as my assignment to say a few words about how the Ford Foundation is responding to issues of sustainability. I think you will see the many ways in which things that are happening at Winrock fit well with the agenda that we have sketched for ourselves.

In September 1991, I had the opportunity to tell our trustees about the work that we were then doing in natural resource management and the environment. It seemed an excellent time to tell the trustees about what we were doing in the environment that had not been underscored a great deal. That way, if they said that's not a good idea, I could say, "well, it wasn't my idea in the first place." Or if they seemed to warm up to it, we could say, "that's fine and we would like to do more." Fortunately, they adopted the latter position.

In doing so, we wanted to think carefully about how the Ford Foundation and its the Rural Poverty and Resources Program could have a particular angle on the environment that would cover topics that were needed and also fit our organizational style and capacity.

We decided that a shorthand way to characterize our interest in the environment was to say that we were interested in the field of the environment and development, and that we were interested in promoting activities that would be helpful as people are trying to invent this new field in which we give balanced attention to protecting our natural resources and environment, while also recognizing the needs that people have for jobs and economic opportunity.

There are three things that we have been trying to do. First, we're interested in finding partners and grantees who we believe are advancing conceptual thinking about environment and development and the methodological tools for knowing when we may have achieved it. Sometimes we do that by helping support individual scholars who may have a sabbatical year and wish to think and write about that. Sometimes we do it by finding university partners such as the International Institute for Ecological Economics at the University of Maryland, a diverse group of people on that campus from both ecology and economics who are trying to think through how to conceptualize something that they are calling "ecological economics."

A second part of our work is supporting what you might think of as demonstrations. A demonstration is where some organization or some group of people is trying to find the answer to an environment and development problem in a particular place at a particular time. So it's a very grounded attempt to advance the field by putting into practice what we know now and then learning from that experience.

I'll share just two experiences. One is an experiment that we and others are helping to fund in the state of Washington where a regional environmental group called Ecotrust is teaming up with the Southshore Bank of Chicago. They have selected an estuary that is important environmentally and that is experiencing stress because of the economic activities that are going on there. Ecotrust and Southshore Bank, which is a very experienced development bank, are putting their heads together with local people to figure out what economic activities and enterprises could be promoted in this location that would create jobs for people as well as being acceptable from an environmental point of view.

A very different example has to do with our work in social forestry. Much of this is going on in Asia. In India, there are some interesting experiments under way that are labeled "joint management." It has to do with finding ways in which local village groups can, in a partnership arrangement, work with the forest bureaucracy to manage forestlands in the vicinity of the villages. They often are, in fact, village lands to which those villages have had long-standing rights to use minor or major forest products. Our interest is to find not only a way to help restore damaged forest resources or to improve them but also to find a way of doing so that creates economic opportunity and economic progress for rural peoples.

A third part of our program is what we're calling "organizational transformation." In talking with CEOs at a number of the environmental organizations, particularly here in the United States, like NRDC, the Environmental Defense Fund, the Nature Conservancy, and the Wilderness Society, we found that many of them had also reached the point where they felt that their agenda in the future should be to do a better job of blending concern with development and a concern with environment. But there are many problems, or at least challenges, in figuring out how to transform one's organization to do that. So, we're in the early stages of working with four or five of the large environmental groups to help them think through what organizational transformations they will require to be able to take on this new agenda, recognizing that it may require new staffing, and new mission statements, but it also may require new ways of relating to one's membership because many of these groups are membership-based organizations. So, there are some real challenges.

You can see then that these three areas—conceptualization, work with demonstration projects, and organizational transformation—are all topics about which Winrock has a good deal of experience and it explains why we find so often points of contact between our programs and yours. We hope that will continue to be the case.

DAN MARTIN
MacArthur Foundation

I'll say a few things about what we're doing with the environment and resources program at MacArthur. Our starting point is the attempt to retard the loss of biological diversity. That necessarily brings us into a lot of concern about sustainable development, including sustainable agriculture. We focus, because of that logic, on a small fraction of the earth's surface because biological diversity is so unevenly distributed. It also leads us to concentrate much of our attention on tropical forests and, to some degree, on coral reefs.

In every priority area, we attempt to support an array of activities because the problems are so systemic. That is, they come from a variety of angles simultaneously, usually including science and policy studies; professional training; general public environmental education, such as public television programs and secondary school programs; traditional conservation activities (which have diminished as they are picked up by the highly unimaginative programs that come from multilateral agencies); and, finally, demonstrations of allegedly sustainable development but only in and around areas of high biological diversity—the places where the greatest urgency occurs.

However little we know about species diversity on this planet, the sense is that we have to take a stab. It's like any other strategic decision, you have to operate on the basis of inadequate evidence.

With that array of work, which gets us into things ranging from comparative studies of forest dynamics to elementary education on environmental subjects, we find that we're increasingly working in the connection between conservation and commerce. This may be just another way of saying sustainable development or the kind of connections the United Nations Conference on Environment and Development attempted to articulate.

One example is the National Institute of Biodiversity in Costa Rica, which has become quite a well-known development. We helped pay for the planning of the institute for several years and then getting it launched with important ties to Cornell University. This became a trendy and stylish operation once it was legitimized by a contract with the Merck Company. The contract in itself is a modest undertaking, but it gave credibility to this kind of operation that has led to vigorous efforts to replicate the model.

Of course, almost every proposal that comes to us claims to be a replicable model, as you can well understand, but this one is being replicated in Mexico, Indonesia, and Brazil and in other countries as well.

Along with that, an interesting connection has developed—that so far is partially hypothetical, but I hope it will be less hypothetical in the future—between the grant-making department and the investment department of the foundation. We found ourselves, to our surprise, making grants for business plans for allegedly sustainable commercial activities. People like me, who have always lived in the nonprofit sector, aren't accustomed to thinking that way, but it's probably the most exciting type of action that we're doing. Once the business plan is in hand, on the basis of some grant support, we find that some of those enterprises come back to the foundation seeking a low-interest loan from our Program-related Investment Department. It's not really an investment. It's a repayable grant, and it works well.

The third step is a real investment, that is to say, a venture capital placement, taking equity in the enterprise. The financial vice-president at MacArthur has set up a special unit in the investment department to work on that.

The potential of that line-up—a grant leading to a low-interest loan, leading to a placement of venture capital, may well be the kind of financing bridge from private-sector activities that will make some sustainable development projects happen. We're optimistic about that.

Another example of conservation and commerce would be our support of efforts to shape markets for products from tropical countries in industrial countries. There are increasing efforts to certify the allegedly sustainable harvesting of lumber. It will be interesting to see how this might work in agriculture—what this would have to do with the application of chemicals and various other kind of treatments that might be made.

Similar to that kind of certification of harvesting techniques is the effort that is emerging now to educate industrial designers and architects about the different properties of various species of tropical hardwoods. Vast areas of the tropical forest are destroyed to get to the few, high-value trees that happen to be the species that the architect in the United States or Japan or Europe has heard about. That's why loggers tear down a lot of valuable trees to get at the few mahogany trees or two or three other species. In the process, they waste magnificent trees with fiber, different physical properties, beautiful grains that can be used in a variety of ways. This is the type of thing that we find ourselves attracted to in helping shape the market that drives what happens in the tropical countries.

We're alson attempting to work with economists to see if there is some potential for the "greening" of academic economics. That's a tough assignment. We think the malignancy of what the World Bank does is a result of the education that economists receive. And let's not mince words, it's malignant. It's not just something that is accidentally not so good and, of course, that leads to other kinds of policy advocacy.

Finally, there is a new kind of trade occurring that I just encountered in Malaysia, which is international trading of carbon-emission offsets. In Sabah, the Sabah Foundation, which owns large timber concessions, now has contracts with New England Power and with a consortium of Dutch utilities to enhance the capacity of their forests to absorb carbon and to improve their harvesting technique to reduce the loss of that carbon-absorbing capacity. There's also a negotiation going on with Southern California Edison to offset the emissions from the Four Corners electric plant in the U.S. Southwest.

In all of these things, rather than seeing economic enterprise as something that somehow conservation must oppose, finding and supporting the creative leadership of people who are out to make money in constructive ways and, perhaps, more sustainable ways for their own business corporations seems to us the wave of the future.

As we find ourselves involved more and more with agricultural activities around those critical forests (the improvement of agriculture where agriculture is appropriate), we find that the work that Winrock is doing is crucial, and that's why it's such a pleasure for me to be with you today.

MICHAEL NORTHROP
Rockefeller Brothers Fund

More and more, I think we will see development assistance shaped by international concerns for climate and biodiversity. The Earth Summit in Rio de Janeiro last year resulted in a climate convention and a biodiversity convention. Those concerns expressed strongly by the international community, I think, will globalize action on several fronts in the development area. Let me suggest examples of what I mean.

Coming directly from the climate convention, we're going to see significant changes in the way assistance is given for energy projects worldwide. The traditional lending work of the multilateral development banks to increase power-generating capacity will have to change. In the future, we just can't have carbon emissions at an increasing scale as we have in the past.

Domestically we're seeing a lot of investment in energy efficiency rather than investing in new energy capability. In the developing world, there's much evidence to suggest that what might have been planned for increasing power generation will be offset by investments in renewable energy sources. I recently was visited by a man from a start-up NGO in Washington who made the comment that if you're 300 meters beyond the grid in a developing country, it is economically cost-effective to have a small solar system in your house. The cell would provide several lights and a television, and would be a huge step up for 2 billion rural dwellers in developing countries.

Second, I think we're going to see, in part because of work of MacArthur and Ford, significant changes in how world assistance programs deal with biodiversity. In the past, we've had lots of money granted and lent for wildlife-protection schemes that have failed, and biodiversity has been eroded as habitat has been eroded. We're seeing examples now worldwide, where we are beginning to manage for ecosystems and habitat. Secretary of Interior Bruce Babbitt in his national biological survey is leading the charge domestically on this. I think internationally we will see the need for institutions like the World Bank to reflect that understanding in the types of assistance that they provide.

Finally, in part because of the tremendous pressure likely to be exerted by concerns of climate and biodiversity, it's likely that we will see change in the profiles of development institutions themselves, both the multilaterals and the bilaterals. They will have to change. The forces being exerted on them are coming from outside the communities that they are used to operating within, namely the economics profession. They have been forced to accept anthropologists, ecologists, urban planners, sociologists, and many other people with noneconomic skill sets as they have gone forward in thinking about how to make loans and grants.

As you think about those three issues—energy, biodiversity, and change in the development banks—as being patterns for future change in development, you see a pattern of cross-disciplinary thinking that will be required in all of those areas. That also will foster change in the World Bank, at USAID, and in several other places. The cross-disciplinary thinking that has been so evident here at Winrock over the last couple of days will be more and more important.

H. PAT PETERSON
U.S. Agency for International Development

I want to address a couple of points that Vernon Ruttan raised. On declining prices of wheat since the 1850s and rice since 1950—some environmentalists would say that one solution to that is that you don't have the "real" price at the time of transaction or consumption. In other words, there are some externalities there that aren't reflected in the market prices. I'm not sure that I agree with that but that's certainly an interpretation.

Vern also commented that the Cold War is over and that we don't have a follow-up. The president no longer can go to Congress and say, "If you don't give me the money, Russia will get you." He's right the Cold War is over. Unfortunately, in terms of my budget, Russia has got me. I don't deny the priority. It is an uncomfortable set of circumstances.

He asked about an encore to the Green Revolution, and that's one of the things that obviously we will have to deal with. Making a solution to food and hunger a major issue for development might be one of the encores. But whatever the encore is, it is going to have to feed about 50 percent more people in the next 20 to 30 years. So we need to crank that into our calculation.

Yesterday I felt fairly good because I had Pierre Crosson telling me not to invest in remediation of environmental problems. Then I had David Seckler say, don't invest in drainage or rainfed agriculture. Then I had Rob Paarlberg saying, don't invest traditionally because that won't empower people. And I thought, by the end of the session I'm going to find something down at the level that I can afford because of all these things I don't have to invest in anymore. Unfortunately today, that wasn't the case. I've been told that there are some things that we need to invest in: forestry, biotechnology, livestock and range management. Also, I was a little bit concerned in the biotechnology discussion when I didn't get a rousing response as to whether or not biotechnology can feed the world because the answer was no, it can't and not only that, Pat, it will cost you more money rather than less money. And I combine that with what Ambassador Blake said yesterday about aid being down the tubes and today that if current trends continue there will be less money for agricultural research. I think probably his assessment is accurate.

Bob Berg noted that there probably will be a shrinkage in the U.S. foreign aid budget, and there will be a number of people who are competing for a larger share of a shrinking pie. That's something that those of us in agriculture have to keep in mind because we probably will have to do the same thing.

It leads to a major problem. Its not clear that there's sufficient investment going into agricultural research to even maintain the flow of technology that we invested so much in to get the Green Revolution. I think we're losing a lot of that around the world. Indeed, I believe that Bob Blake's comment about the CGIAR system being the last bastion of that investment is probably valid.

One conclusion is that either we need to do more research to develop new technology to increase productivity or we need to bring in new lands. New lands don't seem to be there. Therefore the solution has to be in research.

With that there are some hopes: First, in the technology of research, we've talked about the fact that we have computers and communication now that we didn't have that before. Much more rapid communication can take place between scientists around the world, and, it is hoped, with the breakdown of the Soviet empire, we can also communicate with our colleagues behind what was the Iron Curtain. Also, there are some technologies in biotechnology that will allow us to increase the rapidity with which we can get the research results off the laboratory bench and into the farmer's field.

A second conclusion is that we can do much better integration at the field level. We've talked about getting foresters to talk to agriculturalists to talk to irrigation people.

Third, we need to revise the roles of the current institutional actors, the international agricultural research centers, the national agricultural research systems, the private sector, the U.S. universities, as well as the donor organizations, because I think there are some efficiencies there that we can gain.

Finally, we need to work harder to get in a coalition between the environmentalists, the agriculturalists, and the NGOs.

All of those, I believe, are things that Winrock has a capability in helping us with. Winrock is unique in being able to bring the technical agriculturalists, the environmentalists, and the NGOs to the same table. I believe that probably is a legacy that Bob Havener has left because he took three very separate organizations and he now has them working together with a common vision. That legacy probably is going to be a major benefit not only to those of us working in agricultural research but many people in the world.

PETER RIGGS
Rockefeller Brothers Fund

It's interesting that we can talk about a generational change as the catalyst for this conference. And perhaps generation is an artifice, but it's useful in the international arena. We have two strong signposts. The first being Stockholm in 1972 and then last year, the Earth Summit in Rio de Janeiro. So maybe we're under a third generation now, a post-Rio strategy. As part of that, we have to rethink development and resource conservation, not only in areas where we have been working, but in new areas.

There's an interesting link between the two, one geographical and one a program area. Geographically, it's the NIS. In the area of agriculture, I suggest that we give the Ukraine equal time. Often we use Russia as a surrogate for talking about all those countries, but I would hope we get away from that and look at the very different problems that exist in each country.

At the same time, it can't become a theoretical playground and time to try out lots of new theories. We're looking at a part of the world where there are higher expectations and a higher awareness of consumer lifestyles than has ever existed in other parts of the world. The gap has never been larger, and that is dangerous in many ways. It leads to very different kinds of time-phase programs in development because the usual signals are not there. We have an educated populace in these countries that are not able to work in the fields in which they were educated and that is catastrophic. That vacuum can be filled with nationalism, which we've seen in many areas, and it's a time-honored tradition to divert attention away from domestic failures by pushing a nationalist agenda. We're seeing the artificial creation of so-called ancient tensions, which are more or less the legacy of resource abuse over, in some cases, tens of generations. So, that needs to be addressed.

The fact that the NIS has emerged as an area in which we can work is related to a programmatic area, a new focus of attention, which is militarism. We haven't looked at opportunities foregone because of military spending. In Asia we're looking at a potential arms race that in the next few years could negate many advances and certainly tie up billions of dollars that would otherwise be available for development programs and attention to the areas where we critically need to focus our work.

To add to Vern Ruttan's comments about an encore with the Green Revolution, it has to be something that is global, it has to be something that is understandable, it has to be something that engages the public imagination. Wouldn't it be nice if demilitarization could become that encore?

It seems remarkable that the first strong link between environment and militarism came about through the destruction of Iraq and Kuwait. Prior to that time, that link was not generally acknowledged. It was not explicit. More recently, though, I've seen a particularly insidious form of militarism affecting national priorities. That has been in Cambodia, where the various parties to that conflict are financing their military adventures by illegal logging, destruction of various watersheds through gem mining, and abuse of natural resources, which has led to violence and the ever-present threat of violence and continued ecological violence.

We have an opportunity in the NIS to look at the demilitarization of that economy. How do we help the Russians, the Ukrainians, and the Kazakhs, to move their economies from a war footing? Wouldn't it be nice as well if we could apply some of those lessons and bring them home and face up to the fact that our economy has been on a war footing for a long time and, in some ways, it has moved from being at war with the Russians to being at war with our own people? Perhaps we can reimport some of those international experiences in demilitarization into fixing our own economy.

ELISE FIBER SMITH
Winrock International

I think it is important when we talk about what "we" do in agriculture and the environment, it ought to be shaped by the farmer's agenda. We are committed to alleviating poverty in developing countries, especially the poverty of smallholder farmers. We're talking a lot about local control, participation, and decision making by groups of farmers at village levels. We're also talking about how we can look at issues of equity and access, particularly the importance of empowerment at the local level for solving problems of environment and agricultural productivity.

One of the things that we can think about is how can we take our technical capability that deals with significant macro-policy and research agendas and, through NGOs, develop links to local village leadership who can make change happen at the local level. They are the change agents who actually can alleviate rural poverty. The results of policy and research should be to support meeting the needs of the rural poor.

I spend a lot of time looking at the issue of equity, access, and increasing agricultural productivity—and how people can get the food they need and raise their standard of living. But I focus on the participation of women. In Africa, 70 percent of the women are working on agricultural production activities. They are really the people who feed Africa through their small-plot food production. They've received the least credit, the least access to technical resources, and the least amount of services from extension agents, yet they have the a capability to produce a lot more. So, in Winrock's AWLAE program we have looked at where women are in leadership positions in agriculture and the environment—where the gender-sensitive people are that can support this particularly important group of women farmers who contribute so much but have so little access. We discovered that in Africa there are few women trained at the technical and policy level in agriculture and the environment. Winrock is trying to strengthen the leadership preparation and opportunities of women so they are able to contribute to the issues of resource allocation, increased food production, and conservation of the environment. Winrock will continue to expand this effort in the future because we're talking about building leadership at different levels on the development ladder. We have the capability, with the new emphasis on crossdisciplinary efforts, to approach these development problems at in a different way. Everybody knows that overcoming the causes of poverty is really an interdisciplinary set of issues. Finally, I think the policy role is critical for this institution. In this country we have an open agenda. Winrock can help shape the agenda for international development in agriculture and environment. By contributing to that agenda, we can contribute to United States policy in alleviating poverty and hunger in developing countries.

FRANK TUGWELL
Winrock International

I find myself preoccupied, first, with the challenge that we received to define Winrock's niche now that we have a better sense of the landscape of the future with respect to international development and, second, with the problem of responding to what is a dramatically changing donor environment. It's clear that agriculture is not on the agenda of the new administration in Washington, D.C.

There are three drivers that are going to determine the nature of foreign assistance. The first is national security. This goes back to the fact that the Cold War is over. Perceptions of short- and long-term national security are still going to drive decisions about the allocations of resources. What are those concerns? The first is the NIS. The prospect that Russia, Kazakhstan, and Ukraine—countries that hold nuclear weapons—could collapse into anarchy is primary, and we will see money moving there quickly. This will continue to be a problem because they are not going to establish democratic systems and open markets overnight. We're going to have to adapt to that and I think there are ways in which we will.

Second, we will see export promotion and economic competition remain on the agenda as a concern that drives our foreign assistance policy. It's not that we need enemies, but we're going to find that conflicts that before were subordinated to Cold War concerns now becoming more important. International economic conflicts fit that category.

Finally, there is the environment. It is interesting that we have a conflict going on in the administration about where the environment fits. I think the winners will be those who can claim that a host of environmental issues are also security issues—that the air and the water that we share, that the species that we will all rely upon—will all be denied to us or will be damaged in some way unless we take action.

Those are the three main drivers or concerns: national security, export promotion, and environment. Unfortunately agriculture and the long-term productivity increases that we realize must occur are not perceived as a short-term or even a medium-term concerns. So, that's a challenge both for Winrock in terms of its strategy and also the international development community.

I think there are ways in which Winrock can and will respond to that. Already we have more than half of our portfolio in the field of environment. This includes things like forestry, renewable energy, environmental policy. I think we will respond constructively in those areas.

One other theme that has been of interest to me is whether there will be an encore to the Green Revolution. The international development community hit a home run once and the question is, can we hit another? I have feel that it's not going to be technical in the same sense that the first home run was perceived to be (and I think that's partly perception). If there's going to be a new home run, it will be in the field of organization and policy. In his paper Rob Paarlberg connected empowerment, the creation of groups, the movement of real resources to local people and to women as a mechanism by which you can get economic growth and by which you can enhance the environment. Those are organizational and policy challenges, and if we have a home run, this is the area where we'll find it, I expect.

Closing Address

EARL KELLOGG
Winrock International

Among the themes that have been discussed in this conference, there are a number that bear repeating. One development need is to help develop support for, and to nurture, national and sub-national institutions. The pressure for institutional innovation is great in these times and that involves both institutions that work on agricultural research and on extension. While international institutions require assistance, I believe we should focus more on national institutions as well as the NGOs that we've talked a great deal about. These institutional innovations certainly must be sustainable. We have to think seriously about what that means. Certainly there is a need to have national ownership. That will be tricky to do. I think it is an important task in front of us.

NGOs are indeed an important part of this, but we need to be careful in what we ask them to do. We have talked about NGOs being advocates, being institutions to help empower people, being development agencies, being technically based, about achieving increased democratization through NGOs, and about taking control of natural resource bases and systems. We want to load everything that should be done onto those institutions. I suspect that that's probably one of the quickest ways to overwhelm them. I think they do have a role. We have to think carefully about what that is and about what Winrock's role with respect to them is.

I was impressed as well about the number of questions of fact that seemed to be with us yet—soil erosion and degradation, water problems, desertification, biodiversity. Knowledge about the extent to which these are problems and why they are problems is more uncertain than I had expected. It seems to me we heard a lot about what we don't know with respect to the state of these kinds of topics and I was a bit surprised by that.

I want to voice a bit of urgency here as we think about the future. We're adding more people per year than we ever have—93 million people every year. Our population growth rates are falling, but the stock to which we're applying that rate is higher than ever. The demand for food is growing more rapidly than ever. With larger numbers of people, with income increases applied to larger numbers of people, with increased urbanization, and with increased demand for animal products in people's diets, food demand growth can be high indeed. At the same time we have water concerns, decreasing investments in irrigation, decreasing investments in agricultural research, and increasing pressure to be concerned about the environmental shortcomings of current agricultural practices. Yet agriculture is not simply becoming a lower priority, it is dropping off many priority lists altogether. We do have an urgency about all of this that I think we need to recapture in our thinking and in our actions.

We've also been told and helped to understand that poverty is not the sole problem that causes natural resource degradation, but certainly the opposite may be true. In fact, the solutions for food, fiber, forestry, and water conservation are probably still linked to the small holder in one way or another.

Economic growth is not an answer that can overcome all of these problems, but it is an important variable that can contribute to finding solutions. I think we must pay attention to overall growth as well as to equity and environmental concerns. We tend, in development, to swing to one thing and then back to the other. We had a heavy emphasis on growth at some times, and we forgot about important equity and environmental concerns. Then we had a heavy emphasis on environmental issues, and we forgot about the importance of overall growth. Somehow an institution like Winrock and institutions that many of you represent have to keep us from going too far one way and the other in our current thinking.

I want to say a few words in closing about Winrock's vision. Our mission is to alleviate poverty and hunger through sustainable agriculture and rural development. We believe that our contribution should be in developing longer term solutions to poverty and hunger, not only in developing countries in the world but in the United States, in the mid-South, and in Arkansas. Many people here are motivated to do this because there is a basic human dignity that should be present in all people. People do deserve to eat well, they do deserve reasonable health care, they do deserve to have an education, and they do deserve to have a future. Many of us here are committed to this institution because we believe those things strongly. We believe we can complete our mission through human resource development, institutional development, policy development, and technology development and transfer. Broad-based development, helping poor people better secure their future, is important to having a more peaceful and just world. Our role, we believe, is a limited one and we're entering into a time when we're going to be doing some strategic thinking about what that role will be in this rapidly changing context. But we do believe it's important that we do development.

We believe that we are an important institution for bridging the gaps between environment and agriculture and trying to make more sense out of our role as an NGO.

We want to thank all of you for coming and, more important, for working with us to achieve these important missions and visions. Many of you have asked, "What are we going to do when Bob Havener leaves?" And I will say to you that he's only leaving a little bit because we will be using Bob's services as we continue forward in this important time in this institution. So your legacy, Bob, is one that's exemplified by the facts that these people are here in your honor and that this institution is strong, healthy, and looking forward rather than back.

Participants and Authors

An International Symposium in Honor of Robert D. Havener

World Agricultural Resources in the 21st Century: Environmental Quality, Natural Resources, Technologies

Pierre Antoine
Director
Africa and Middle East Division
Winrock International

Mary Barakat
Community Relations Manager
International Relations
Pioneer Hi-Bred International

Sandra Batie*
Professor
Department of Agricultural Economics
Virginia Polytechnic Institute and State
  University

William Bentley
Senior Program Officer
Center for Institutional and Human
  Resource Development
Winrock International

Robert Berg
President
International Development Conference

Robert O. Blake
Chairman
Committee on Agricultural Sustainability
  for Developing Countries

Steven A. Breth
Program Officer
Communications
Winrock International

Paul Brown
Project Coordinator
U.S. Division
Winrock International

Marion Burton*
Attorney at Law

Fee Busby
Director
U.S. Division
Winrock International

Colin Campbell*
President
Rockefeller Brothers Fund

Ronnie Coffman*
Associate Dean of Research and Director
Agricultural Experiment Station
Cornell University

E. Walter Coward, Jr.
Director
Rural Poverty and Resources Program
Ford Foundation

Pierre Crosson
Senior Fellow
Resources for the Future

John De Boer
Senior Program Officer
Asia Division
Winrock International

William Dietel*
Chairman
American Public Radio

Byron T. Edwards*
Investor/Farmer

Jim Ellis
Director
Center for Environment and Sustainable
  Agriculture
Winrock International

Note: Affiliations indicated are for May 1993 when the symposium was held.

Walter Falcon*
Director
Institute for International Studies
Stanford University

Charles Fultz
State Soil Scientist
U.S. Soil Conservation Service

Neva R. Goodwin*
Co-Director
Global Development and Environmental Institute
Tufts University

John C. Gordon
Pinchot Professor of Forestry
Yale University

Thurman Grove
Assistant Dean
College of Agricultural and Life Sciences
North Carolina State University

Lowell Hardin*
Emeritus Professor of Agricultural Economics
Purdue University

Richard Harwood
C. S. Mott Foundation Professor of Sustainable Agriculture
Department of Crop and Soil Science
Michigan State University

Charles R. Hatch
Chief of Party
Pakistan Forestry Planning and Development Project II
Winrock International

Robert D. Havener*
President
Winrock International

Kenzo Hemmi*
Professor of International Economics
Toyo Eiwa Women's University

Robert Horsch
Crop Transformation Director
The Agricultural Group
Monsanto Company

Clive James
Chairman
International Service for the Acquisition of Agri-Biotech Applications

Catherine Jewsbury
Senior Project Economist
Environmental Policy and Training Project
Winrock International

Avtar Kaul
Senior Program Officer
Asia Division
Winrock International

Earl Kellogg
Senior Vice President
Winrock International

Jim Maner
Director
Latin America and Caribbean Division
Winrock International

Dan Martin
Director
World Environment and Resources Program
The MacArthur Foundation

Sandra Miller
Project Director
Arkansas Rural Enterprise Center
Winrock International

Michael Northrop
Program Officer
Rockefeller Brothers Fund

Fanny Nyaribo-Roberts
Agricultural Economist
College of Agriculture and Home Economics
Washington State University

Robert Paarlberg*
Professor
Department of Political Science
Wellesley College

N. S. Peabody III
Chief of Party
Environmental Policy and Training
  Project
Winrock International

H. Pat Peterson
Director
Bureau of Research and
  Development/Agriculture
U. S. Agency for International
  Development

Roberto Rapera
Chief of Party
Philippines Natural Resource
  Management Project
Winrock International

Ned Raun
Senior Associate
Winrock International

Peter Riggs
Program Officer
Rockefeller Brothers Fund

Vernon W. Ruttan
Regents Professor of Agricultural
  Economics
Department of Agricultural and Applied
  Economics
University of Minnesota

David Seckler
Director
Center for Economic Policy Studies
Winrock International

Robert Shults[*]
Attorney at Law
Shults, Ray & Kurrus

Elise Fiber Smith
Project Director
African Women Leadership in
  Agriculture and the Environment
Winrock International

Carol Stoney
Agroforestry Specialist
Asia Division
Winrock International

Jonathan Taylor[*]
Chief Executive
Booker Companies

Frank Tugwell
Vice President
Programs and Global Projects
Winrock International

Jim Wimberly
Program Officer
U.S. Division
Winrock International

[*] Present or former member of the Board of Directors, Winrock International.